Frothers, Bubbles and Flotation

A Survey of Flotation Milling in the Twentieth-Century Metals Industry

Dawn Bunyak

1998

1998, National Park Service, Intermountain Support Office, Denver, Colorado

First printing

The Shenandoah-Dives (Mayflower) Mill,
Silverton, Colorado in the 1930s.

The ore arrived from the mine uphill from the mill via tramway buckets at the arrival
terminal located on the right. From the terminal, the ore moved by conveyor to the
top of the mill building where it was crushed and ground for the flotation process.
At the completion of the flotation process, the concentrate arrived at the ore bins
in the lowest building level. The mill office building is to the left of the arrival
terminal.

Table of Contents

List of Illustrations

Photographs: (unless noted, photos are property of Dawn Bunyak)

Figures

In May 1995, Dawn Bunyak, a graduate history student at the University of Colorado-Denver, contacted the National Park Service in Denver with an offer to work as a volunteer. Dawn was looking for a research project in history or historic preservation. Serendipitously, the San Juan County Historical Society had recently contacted our office asking for help in preparing a National Register nomination for its historic Shenandoah-Dives Mill, located in Silverton, Colorado. It turned out to be a very good match. And, for the next year and a half, Dawn worked with the historical society and the National Park Service to prepare that nomination. On April 3, 1997, the Shenandoah-Dives Mill was listed on the National Register of Historic Places, as part of the expanded Silverton Historic District.

However, Dawn's research into the history of the Shenandoah-Dives Mill raised as many questions as it answered. While working on the project, Dawn had discovered that there were virtually no published histories of selective flotation milling. In addition, there had never been a survey of extant flotation mills in the United States. How many other flotation mills were there? What kind of shape were they in? The Shenandoah Dives appeared to have an exceptional level of integrity, as all of its machinery was intact. But, without knowing the number and condition of other such mills, there could be no comparative analysis. Moreover, although very little was known about selective flotation mills, there were indications that they were becoming a vanishing resource. Through demolition and neglect, an important part of our industrial heritage appeared to be slipping away.

In the fall of 1996, Dawn proposed a project to: 1) research the history and technology of selective flotation milling; and 2) identify extant flotation mills throughout the United States. The National Park Service, in cooperation with the University of Colorado-Denver, funded Dawn's proposal. This report is the result of that project, and its publication was also funded by the National Park Service. In the course of her research, Dawn had interviewed numerous federal and state agencies – as well as private mining and milling companies – who own historic mills. Several of those agency representatives and owners shared Dawn's frustration at the lack of a historical context for flotation milling, and the scarcity of information on other such properties. This report helps fill that gap, and is an important step in helping identify a significant component of our Nation's mining and milling history.

Lysa Wegman-French and Christine Whitacre, Historians
Cultural Resources and National Register Program Services
Intermountain Support Office-Denver
National Park Service

Abstract

Flotation was the most important development of the twentieth century for the mining industry. It is an innovative processing method for separating valuable minerals locked in waste rock, and abandons basic principles of ore treatment by gravity methods previously followed by metallurgists. At the end of the nineteenth century, the world had almost exhausted its supply of high-grade ore. Mining engineers and inventors worked at a feverish pitch to create a concentration method to treat the world's abundant supply of low-grade and complex ores, which were often difficult to extract from waste rock. In 1905, the world's first commercially successful flotation mill was established at Broken Hill, Australia. Soon, experimental plants in Australia, Great Britain and the United States also spread the news of their success with flotation. Almost overnight, the mining industry shifted from exploiting diminishing high-grade ores to an abundant supply of low-grade ore.

Flotation propelled the mining industry into a new age. It can be argued that flotation is the beginning of modern mining in the base metals industry. While earlier concentration methods, such as gravity followed by cyanidation, were effective in processing gold ore, flotation opened the minerals market. Metal production was increased to 24 metallic and 19 non-metallic minerals by the mid-twentieth century. Flotation can be used in processing metallic ores (e.g. copper, lead, zinc, gold, silver) or non-metallic ores (e.g. clay, phosphate, coal). The most widely used process in the world for extracting minerals, flotation also recovers several metals from a single ore body into two or more concentrated products. The development of flotation diverted a crisis in industry, as well as allowing increased minerals production in the twentieth century.

Although a very significant component of America's mining industry, flotation mills are also an endangered resource. Twentieth-century flotation mills use the same fundamental method developed at the turn of the century; only the equipment and its size has changed. However, with developments in equipment and increased extraction of ore, the small mills of the early decades have become obsolete. The flotation process, once self-contained in one small building, is now housed in buildings covering acres of land. With this growth, a vital portion of our early mining heritage has been wiped out by removal of early mills for construction of new, larger and more efficient plants. Salvaging of equipment and buildings has reduced the number of mills even more. Natural elements have adversely affected older mill sites that have been abandoned and neglected. Environmental concerns have resulted in wholesale removal of mills, equipment, even the ground they sat on, whether or not they were hazardous sites.

In this study, the author conducted an overview of flotation mills in the United States in order to determine the number and status of extant historic resources. The author found only two fully intact, 500-2,000 ton flotation mills left in the United States: the Shenandoah-Dives Mill in Silverton, Colorado and the Ely Valley Pioche Mill in Pioche, Nevada. These one-to-two thousand-ton per day plants are excellent representations of mills built in the 1930s. Without preservation of these sites, and other increasingly rare vestiges of historic mining and milling sites, America will lose its mining heritage. Modern technology and mankind's neglect are wiping out America's mining past.

Introduction

At the 1989 Historic Mining Conference in Death Valley National Monument, mining historian Robert Spude [then Chief, National Preservation Program, Rocky Mountain Region, National Park Service] discussed the 1920s and 1930s preservation and interpretation of the mining frontier from the 1849 California Gold Rush through the end of the nineteenth century. Romanticized tales about the miner and mining communities built around nineteenth-century gold strikes resulted in a preservation movement of the architectural relics from that era's mining community. However, the industry that created the mining towns and eventually, cities, was not interpreted or preserved. Moreover, the mines and mills that represented early and mid-twentieth century technology have been virtually ignored. The remnants of mines, mining camps, and mineral processing plants fell into disrepair and neglect, eventually removed as eye sores in the community. Thus, the economic activity that initially created the mining West was removed piece-by-piece, as machinery was scrapped; token pieces of equipment were hauled away to museum exhibits; and headframes, mills, and smelters were bulldozed into the ground.

One of the most important properties associated with mining is the flotation mill. The development of flotation opened new mineral resources for the mining industry and averted a twentieth-century industrial crisis with the exhaustion of high-grade ore supplies. Disregarding earlier (and ineffective) specific gravity methods, inventors developed flotation, an ingenious process separating valuable minerals from complex ore bodies. Basically, flotation "floats" the minerals on a waterbath, where it can be collected and dried in the mill, then sent to smelters for final processing. Flotation, considered the beginning of modern mining, opened up the world's low-grade mineral deposits for large-scale commercial use by increasing recovery and the grade of concentrates.

The purpose of this report is to provide a thematic framework for identifying and preserving a significant component of America's mining and milling history: twentieth-century flotation mills. This report has been written to establish a national context for minerals flotation milling in 500-2,000 ton industrial plants. Though small by today's standards, historically these were considered to be large plants. The time frame of this report is 1905 through 1959. The year 1905 marks the development of flotation for commercial use at Broken Hill, Australia. The 1950s marks a turning point in the mining industry toward open-pit mines and the resulting sudden enlargement of the size of flotation mills. In addition, this report gives an overview of the history of early twentieth-century flotation milling, identifies extant historic flotation mills, and emphasizes the importance of saving the vestiges of remaining historical properties.

In 1995, the San Juan County Historical Society, located in the National Historic Landmark (NHL) community of Silverton, Colorado, acquired several acres of property including the Shenandoah-Dives Mill Complex, locally referred to as the Mayflower Mill. Constructed in 1929, the mill was an outstanding example of selective flotation milling. Operational until 1992, the mill's machinery was in place. The society approached the National Park Service (NPS) in Lakewood, Colorado for assistance in listing the mill on the National Register of Historic Places. As a graduate student at the University of Colorado at Denver, I was offered the project under the direction of Christine Whitacre and Lysa Wegman-French, historians for the Rocky Mountain System Support Office. On April 3, 1997, the mill site was listed as a contributing property on the National Register of Historic Properties, enabling the town to obtain grants and monies for the preservation of the mill complex as an interpretive center for mining and milling in the Rocky Mountain West.

The Colorado State Historical Society Board and the NPS then encouraged Silverton to complete the research necessary to determine if the Shenandoah-Dives Mill would also qualify as a contributing feature in the NHL district. With the cooperation of Silverton and the NPS–under the auspices of a cooperative agreement with the University of Colorado at Denver- I prepared the following report and case study which places the Shenandoah-Dives Mill within the national context of flotation mills. Also included within is a listing of all the extant flotation mills that I was able to locate through my research. In this report, I will discuss the methodology of my research pertaining to early twentieth-century hard-rock milling operations; a brief history of the milling industry, including the history of the Shenandoah-Dives (Mayflower) Mill; the listing of extant mills; and final conclusions and recommendations.

Methodology

The goals of this report are: 1) to develop a historic context for flotation mills used in the metals industry in the twentieth century and 2) to identify extant 500-2,000 ton-per-day mill sites in the United States. In order to develop this context, this report will be divided into several sections: methodology, overview of the mining industry, overview of ore processing, the history and evolution of flotation, defining and evaluating property types, and results of this study.

The time frame of this report begins with the development of flotation for commercial use in 1905 through 1959. Developments in flotation technology included the creation of larger crushing machines (that could crush ore finer or crush larger quantities than earlier machines) and the development of cells for flotation. Otherwise, flotation processes have remained fundamentally the same from their appearance on the commercial scene in Australia. The 1950s marks a turning point in the mining industry towards open-pit mines. In addition, the 1959 ending date was derived from the National Register of Historic Places requirement for properties achieving historical significance at least approximately 50 years ago.

Due to the scope of this study, it was not possible to visit all of the mills. I have relied upon the assistance of experts in the mining world for descriptions and evaluations of integrity for some of these sites.

It is, also, important to realize that this report is dealing principally with the engineering aspects of milling, not the architectural aspects of the building, nor other related contexts such as labor history.

Mills can also be referred to as processing plants, reduction works, and concentrators. For purposes of this report, "mill" refers to a "flotation mill" unless otherwise noted.

State Historic Preservation Offices

In order to develop the historic context of flotation mills in the twentieth century, I contacted via electronic mail the State Historic Preservation Officers of each state. I also called individual State Historic Preservation Offices to inquire 1) if the state had developed a mining historic context, 2) if they had a listing of experts in their state's mining history, and 3) to obtain information on the National Register-listed mining and milling properties in their states.

Very few states have developed a historic mining context to assist historians and researchers in mining history. States that developed contexts include Minnesota (Minnesota's Iron Ore Industry); South Dakota (Black Hills Mining Resources Historic Context); and Arizona (Gold and Silver Mining in Arizona). Pennsylvania has a context on coal and coke in the state's bituminous regions, and is currently preparing a study on the anthracite region. Colorado's mining contexts can be found in the state contexts for engineering (Industrial-Smelting and Mining-Stone Quarrying) and mountains (Gold Rush and the Placer Mining Frontier 1858-1868, Precious Metal Mining as an Industry 1860-1920, Technology of Mining in Colorado's Mountains 1859-1900, and Lead, Zinc, and Other Mining 1860-1945). Montana's Department of Environmental Quality has developed an inventory of their historic mining properties, extant and in ruins. Several of the states gave names and numbers for experts in mining history in their states (refer to section on interviews). Staff members of various Preservation Offices reviewed their files on mining properties listed on the National Register of Historic Places to determine whether such sites contained a mill or mineral processing facility (refer to following section for more information).

National Register of Historic Places/National Historic Landmarks

The National Park Service prepared a national mining theme study, "A Mining Frontier Context," in 1959. This National Historic Landmark survey focused on the towns and camps associated with nineteenth-century mining. "A Mining Frontier Context" helped prompt preservation of nineteenth-century mining camps and towns. However, the context did not cover the technology and economic activity of the mines, mills, smelters, or the American mining landscape. The American mining landscape includes signs of the extraction process (mine openings, headframes, adits); milling process (mills, aerial trams, railtracks); and disposal of waste (tailing heaps and ponds). Indeed, mines and mills were responsible for establishing the mining camps, towns and cities that populated vast regions of the United States. Another government publication, National Register Bulletin No. 42, Guidelines for Identifying, Evaluating, and Registering Historic Mining Properties, offers excellent guidelines in surveying and registering properties with the National Register of Historic Places.

The electronic database for the National Register of Historic Places was surveyed to compile a listing of mining operations. In addition, I consulted "Appendix B: Historic Mining Properties in the Records of the National Register of Historic Places" in Death Valley to Deadwood from the proceedings of the 1989 Historic Mining Conference in Death Valley

National Monument.[1] Those states that have properties directly related to historic mining in the United States include: Alaska, Arizona, California, Colorado, Georgia, Idaho, Illinois, Iowa, Massachusetts, Michigan, Minnesota, Montana, Nevada, New Jersey, New Mexico, Nevada, North Carolina, Oklahoma, Oregon, South Carolina, South Dakota, Tennessee, Texas, Utah, Washington, Wisconsin and Wyoming. Of those states and properties, there is only one site with a fully equipped historic flotation mill listed on the National Register: the Shenandoah-Dives Mill in Silverton, Colorado.

Historic American Buildings Survey and Historic American Engineering Record

In the index of the Historic American Buildings Survey (HABS) and Historic American Engineering Record (HAER), I found documentation filed for reduction works, concentrators, smelters and mills. Many of the properties were documented as a result of mitigation due to threat of removal. The rest of the buildings are either ruins, foundations, or sites. HAER documentation includes these remaining sites and ruins: four deteriorating flotation mill sites and two sites with foundations and historic archeological remains.[2]

Primary Sources

There are no current surveys covering twentieth-century mining and milling. An important source of information was the contemporary mining and mineral journals of the 1920s and 1930s. Experts who had published journal articles also wrote texts on ore dressing, mining and general handbooks.

Secondary Sources

The first difficulty encountered was finding secondary materials on the history of mining and milling after the turn of the century. Articles, texts and handbooks have been written about every two decades from 1900-1960 chronicling flotation and technological advances in the industry. Important reference sources were the classic mining engineering handbooks: Robert Peele's, Mining Engineers' Handbook; Robert Richards' texts, Ore Dressing and Textbook of Ore Dressing; and T. A. Rickard's histories about mining and flotation processes, A History of American Mining, Concentration by Flotation, and Flotation. Modern works include government documents, e.g. Principal Gold-Producing Districts of the United States, Paul Thrush's 1968 dictionary of mining and mineral terms, and Duane Smith's histories of mining in the West including Mining America: The Industry and the Environment, 1800-1980. However, a survey of the history of flotation milling from the 1960s to the present remains to be written. There are technical manuals on a diversity of topics within mining and milling for those who are pursuing an occupation in mining. Definitive works of the mineral industry, including the processes of mining, milling, smelting, and transporting of concentrates to manufacturing centers, are rare. Cedric E. Gregory, an Australian miner and expert in the field, wrote an introductory textbook, A Concise History of Mining, for universities. The text format and language is beneficial for those looking for an overview on mining. In addition, an anniversary volume, including the development and process of flotation from 1911-1961, Froth Flotation 50th Anniversary Volume, D. W. Fuerstenau, editor, is helpful in outlining froth flotation. Milling Methods in the Americas, edited by Nathaniel Arbiter, published in 1964 in conjunction with the Mineral Processing Congress is helpful in bridging early flotation mills to the mills of the 1960s. Research for primary and secondary materials was done in The Colorado School of Mines Arthur Lakes Library; Denver (CO) Public Library, Western History Department; Auraria Library, University of Colorado at Denver (CO); and the CARL link to library catalogs.

Electronic Sources

A request for information on historic mining and milling was posted on the internet list serve for the Mining History Association. Responses came from several experts concerning ruins and sites in their states, as well as bibliographic recommendations.

Interviews

In order to find and evaluate the significance of extant flotation mills, it was necessary to contact mining experts in each of the principal mining states. Queries were made concerning existing flotation mills and their condition in their respective states. As Robert Richards pointed out in his 1940 textbook, Textbook of Ore Dressing, "Over a period of twenty-five years or even less mills undergo numerous changes and many of them become so modified as to be unrecognizable."[3] His words are substantiated in the state of mills today. In addition, older mills have often been destroyed to make way for larger facilities.

To familiarize myself with the history of mills, I drew up a partial list of mills known to be in operation in the twentieth century. With my list (drawn from journals and books), I began the long process of contacting miners, mining engineers, geologists, mine officials, agencies, and mining property owners themselves. National, state and local officials in historically recognized mining areas of the United States were contacted.[4] Representatives of state mining associations (AZ, UT, WY, CO,); state historical societies; professors in states' schools of mines (CO, MI, MT, NM, NV); Bureau of Mines and Minerals (NM); Geological Survey Offices (ID, MT, WA, WY); Bureau of Land Management (CA, NC); mining and mineral museums (AZ, MT, SD, MN); Department of Mines and Mineral Resources (AZ); Department of Natural Resources; and Departments of Environmental Concerns (MT), to name a few, were interviewed about the history of early twentieth-century milling operations in their respective states. In all, twenty-five states with mining histories and over forty individuals recognized as experts in their fields were contacted for information on mining within their state borders.

Endnotes

1. National Park Service, Division of National Register Programs, <u>Death Valley to Deadwood Kennecott to Cripple Creek: Proceedings of the Historic Mining Conference January 23-27, 1989 in Death Valley Monument</u>, ed. Leo R. Barker and Ann E. Huston (San Francisco: Government Printing Office, 1990), Appendix B, p. 174-185.

2. Two mill sites, Silver King (HAER No. UT-22-B) and Keystone (HAER No.UT-28), in Utah have HAER documentation, but it is not clear from the evidence gathered whether the mills are flotation or not. The sites are: Mary Murphy Mining and Milling Complex, Iron City, CO; Comet Town Site, Basin, MT; Slowey Mill at Nancy Lee Mines, Superior, MT; Royal Basin Mine and Mill Site, Maxville, MT; Tintic Standard Reduction Mill, near Park City, UT; and Mill Pond (a lead, selective flotation mill), Grandin, MO.

3. Robert Richards, <u>Textbook of Ore Dressing</u>, (New York and London: McGraw-Hill Book Co., Inc., 1940), 355.

4. I used Richard Francaviglia's outline of U.S. historic mining areas found in <u>Hard Places: Reading the Landscape of America's Historic Mining Districts</u> (Iowa City, IA: University of Iowa Press, 1991), 6-7.

Chapter 1 - Historical Overview of the Mining Industry

American mining encouraged the exploration, expansion and settlement of vast regions. From the first workings of mineral deposits by the Native Americans to the open-pit mining of the 1990s, mining has left its mark on the face of the land. Mining is the extraction of ore from the surface and sub-surface of the earth. Ore can be defined as a natural mineral compound of elements which have at least one metal element. The first miners hand picked ores, often referred to as native ores, visible with the naked eye. Later, miners realized that coarser minerals not rock size could be gathered from streams and creeks by simple gravitational concentration in water, with heavier metals settling to the bottom of pans or rockers (a crude box rocked with one hand[1]). When man began to dig into the earth to retrieve minerals, technological advancements in excavation and supports thrust underground mining to the forefront of the industry.

Gold Rush Era

The 1849 gold rush launched the United States into a period of western exploration, expansion, and settlement. The dream of striking it rich prompted many hardy souls to try their hand at mining. The West offered new opportunities for those who could and would brave the loneliness and hardships encountered in its isolated regions. Mining was "low tech" using hand tools, such as pans, rockers, and sluices (a shallow trench or trough with a riffled bottom for collecting heavier minerals like gold), and water. By the 1860s, mining spurred settlement from the Rocky Mountains to the shores of the Pacific. The gold rush in California was succeeded by gold rushes in many of the Rocky Mountain states. Settlers came seeking gold, land, employment, religious autonomy, freedom or a new beginning. For some, their dreams were fulfilled. Others rode the crest of wealth and poverty with the rising and falling tide of the mining industry.

After the Rush

After the Civil War, explosive growth in big business, fueled by Eastern entrepreneurs, changed the face of mining in the West. By the late nineteenth century, mining had evolved from the single miner with his pack horse to entrepreneurs and syndicates backing large-scale mining entities. Geologists located probable mineral veins that could produce enough of the desired mineral or metal to justify the high cost of extraction. Most precious metals, e.g. gold and silver, were buried deep within the mountains and valleys of the West. High costs of extraction eventually restricted mining to large companies, who could afford to invest in equipment to open mine shafts and tunnels to bring out the ore. Excavating and extraction involved drilling, blasting, hoisting, and hauling. After extraction, the ore was processed in the benefication stage. Benefication includes crushing, concentration (e.g., gravity, flotation, amalgamation, leaching), and smelting. These processing steps will be discussed in the following section.

Improved transportation also advanced the state of mining. Railroads crisscrossed the western region allowing for the movement of ore to smelters in industrial centers throughout the United States. Crews of railroad laborers enticed by dreams of finding gold stayed in western regions to mine. Miners and mine owners continued to be affected by the ever constant rise and fall of the market even with investments from Eastern industrialists. The mining industry suffered a major decline after 1893, due to a combination of factors including changes in monetary standards and the near depletion of high-grade ore.

Twentieth-Century Mining

With the imminent depletion of high-grade ore at the end of the nineteenth century, the mining industry faced a severe crisis. The development of the cyanide process ensured that low-grade gold ores could be treated, but the mining industry had come to depend on minerals other than gold. The development of flotation in the twentieth century ensured the future of mining and moved it into a new phase, a base-metal market for copper, lead, zinc, and iron, with gold and silver as a by-product. The minerals market took off and remained relatively constant until 1929.

After the 1929 stock-market panic, the mining and minerals industry experienced one of its darkest and longest economic depressions. New mining and milling ventures were halted. Investors and syndicates were fighting to survive in a nation-wide depression and were not interested in any speculative deals. Mines closed and miners left in droves to find work wherever it could be found. Mills and smelters were moth-balled. Operations continued only in the largest industrial centers that could financially weather the down turn in the market. Small operators lost their businesses.[2]

By 1933, with the rest of the world moving towards a second World War, the minerals industry began to pull out of its slump. European markets created a demand for armament and gold reserves. Many mines reopened in 1934 when President Franklin Delano Roosevelt nationalized silver and called for the federal government's custody of all gold. Silver and gold prices jumped. Gold prices increased from its 1934 value of $20.67 per ounce to a 1937 value of $35 per ounce. A flood of miners returned to mining regions across the country.[3]

Whereas base metals and by-product gold production from 1907-1943 had contributed to the precious minerals industry, base-metals production of copper, lead, zinc, iron, and molybdenum, literally took over the industry during World War II. A surge in gold mining, caused by the federal government's call for that metal during the depression, was swiftly curtailed by the War Production Board's Order L-208 in the spring of 1942. The order closed all mining operations where gold contributed more than 30% of annual production income. Base-metal companies supplied 4,869,949 ounces of by-product gold in 1940. As a result of Order L-208, the production of by-product gold ore was reduced to a combined total of one million ounces for the years 1944 and 45.[4] The emphasis now was on essential minerals necessary for the production of armament.

The federal government set controls on the mining industry during the war. Congress gave various committees and agencies, e.g. the Defense Logistics Agency, within the federal government the right to award contracts and allocate scarce materials. Goods were deemed essential or non-essential by the War Production Board (WPB) under the leadership of Donald M. Nelson. Nelson and the WPB had authority to commandeer materials, assign priorities, convert and expand plants and bar non-essential production. Defense contractors were guaranteed funds to cover costs, including a hefty profit.[5] This artificially-induced economic market would come back to haunt the mining industry at the end of the war.

Companies, such as the Shenandoah-Dives Mining Company in Silverton, Colorado, quickly contacted Donald M. Nelson and the WPB concerning the production of base metals in their mining operations. Backers of such enterprises, including the leading mining periodical, Mining World, wrote letters supporting company reports and asserting that several Western metal producers must be given preferential ratings due to the importance of their essential metals production.[6] In the end, base-metal mines were vitally important for their contribution to the war cause.

Scrap Drives

Much of the destruction of early twentieth-century mining properties began with the iron scrap drives of the 1940s.[7] As the Second World War waged around the world, markets were interrupted. Shortages of rubber, metals and other goods led to scrap drives. The Office of Civilian Defense led scavenger hunts for the collection of scrap materials.[8] Slogans calling for citizens to "Hit Hitler with the Junk" and songs, including crooner Bing Crosby's, "Junk Will Win the War," encouraged collection of scrap metal. One report stated that a phenomenal five million tons of scrap were collected in three weeks.[9] In Wyoming, a twenty-ton steam engine was taken apart and hauled in to a collection center. Eager participants built a new road several miles long just to haul scrap metal from the site of the engine to the collection center. Parts from the USS Maine were dredged out of Havana Harbor. The town of Peetz, Colorado, with a population of 207 gathered up 225 tons of scrap.[10] Abandoned

mines and mills were not exempt from these scrap drives. If caretakers were not in charge of a property, equipment was removed and taken to collection centers.

After World War II

After the war, the mining industry was plagued by rising costs. The U. S. government had set a fixed selling price upon metals to cap growth during the war years. Federally subsidized mining and milling of minerals, such as chromite, manganese and lead was abruptly withdrawn at the conclusion of the war. Companies folded when the market for their product dried up. Still laborers demanded higher wages after holding back their demands during the war.[11] While other industries flourished, base-metal mining languished under high costs of labor and diminishing markets.[12] Between 1947 and 1951, a demand for lead and zinc in many industries temporarily fueled the U. S. mining economy. However, by 1954, small companies in the mining industry were well into a bust period. Industry wide closures of these companies lasted into the 1960s.[13] Charles Chase, General Manager of the Shenandoah-Dives Mining Co., wrote Congressman Wayne Aspinall in May of 1953 seeking government intervention through loans to mining enterprises.

He reminded the congressman of the artificial market Congress created during the war,

> To be noted especially is the fact that the economic climate of recent years has been artificial. In the war years, wages were raised by government to help employees. The Company, being unable to pay more, was given "premium" prices for metal. The war ended and premium prices ended, but advanced pay scales remained. Industry-wide pay scales, easily honored by the industrial giants, are beyond reasonable reach of industrial pygmies, but must be used.[14]

Despite a letter campaign, Shenandoah-Dives Mine and Mill was unable to obtain funding and, like many smaller industrial operations, was forced to close its operations in1953.

Mining in Post World War II Era

In the 1965 Minerals Yearbook published by the U. S. Bureau of Mines, authors Koschmann and Bergendahl noted that of the twenty-five recognized mining districts: eight were dormant, five produced less than one hundred gold ounces annually, and twelve maintained activity comparable to prewar periods. With the demise of small mining and industrial operations, giants in the minerals industry (e.g.

ASARCO, Anaconda, Kennecott and U.S. Steel) bought defunct mines, mills and mineral rights. By the 1970s, heavy equipment, open-pit mines, and large mills and smelters processing sixty to one hundred thousand tons of ore a day became the norm of the mining industry.[15]

Endnotes

1. U. S. Department of the Interior, National Park Service, The National Survey of Historic Sites and Buildings: The Mining Frontier, by Dr. Benjamin F. Gilbert, ed. by William C. Everhart, (Washington D.C.: Government Printing Office, 1959), 16.

2. Carl Ubbelohde, Maxine Benson, and Duane A. Smith, A Colorado History, 6th ed. (Boulder,CO: Pruett Pub. Co., 1988), 294-304; U. S. Geological Survey, A. H. Koschmann and M. H. Bergendahl, Principal Gold-Producing Districts of the U.S. (Washington, D.C.: U.S. Government Printing Office, 1968), 5-6; and Arthur M. Schlesinger, Almanac of American History (NY: Barnes and Noble Books, 1993), 438-459.

3. Principal Gold-Producing Districts, 5-8 and Schlesinger, Almanac, 438-459.

4. Principal Gold-Producing Districts, 5-8.

5. Ronald H. Bailey, The Homefront U.S.A (NJ: Time-Life Books, Inc., 1978), 77-80.

6. "Shenandoah-Dives, Part II," Mining World (June 1942): 10.

7. Dick Graeme, phone interview by Dawn Bunyak, 28 March 1997.

8. Bailey, Homefront, 106-107 and Geoffrey Perrett, Days of Sadness, Years of Triumph: The American

People 1939-1945 (NY: Coward, McCann & Geoghegan, Inc., 1973), 233-234.

9. Perrett, Days of Sadness, 233-234.

10. Ibid.

11. Koschmann and Bergendahl, Principal Gold-Producing Districts, 6; Schlesinger, Almanac, 514; Robin McCulloch, phone interview by Dawn Bunyak, 26 March 1997.

12. William Jones, interview by Dawn Bunyak, Silverton, CO, 15 June 1995; Graeme, interview; and Koschmann and Bergendahl, Principal Gold-Producing Districts, 6-8.

13. Graeme interview.

14. Margaret Barge, "History of the Shenandoah-Dives Mining Company," TMs, 1959? unpublished manuscript, San Juan County Historical Society, Silverton, Colorado, 10.

15. Graeme interview, and McCulloch interview.

Chapter 2 - Ore Processing Industry

Mining involves a series of processes. In the publication, The Archaeology of Mining and Miners: A View from the Silver State, Donald Hardesty suggests that mining technology is a process of extracting minerals of economic value from rock. In his discussion of this process, Hardesty lists several sub-systems, including removal of the ore body, taking it to the mill, crushing the ore, extracting precious metals, and dumping the waste.[1] Delivery, crushing and concentration takes place at a mill. Smelting of concentrates takes place at a smelter on or off a mill property. In the previous section, an overview of mining history was outlined. In this section, discussion centers on the ore processing industry.

Mineralogy

In Gravity Concentration Techniques, Richard Burt and Chris Mills explain, "The key to effective mineral separation is an understanding of the mineralogy of the ore to be treated."[2] Distribution of minerals in the United States is controlled by a complex set of geological circumstances. In terms of mineralogy, the United States can be basically divided into two geological zones: metallic and non-metallic (Fig.1).[3] The geological terrain of the East can be broadly described as a non-metallic mineral base, e.g. crushed stone, sand, gravel, clay, iron and coal. Western geology is broadly metallic in nature, e.g. gold, silver, copper, zinc, tin, and aluminum.[4] The East has soft rock and the West hard rock. Hard rock is simply "hard rock" in that it is difficult to mine and process for its metals.[5] For purposes of this report, minerals and metals will be referred to as metals, since we are primarily discussing the flotation of metals for industrial use.

Ore Processing

After the miner has extracted the ore (a mineral compound with at least one metal element), the next stage is removing valuable minerals from the encasing valueless rock in the milling process. This process is called mineral or ore dressing. Flotation often required crushing and grinding to a very fine powder to achieve liberation of the ore minerals. Technically, ore processing is the mechanical separation and collection of valuable minerals from valueless material of an ore.[6] In order to free valuable metals, the ore is crushed into a sand or a powder in the reduction stage. The crushed ore must then be

Fig. 1—Historic Mining Areas of the United States

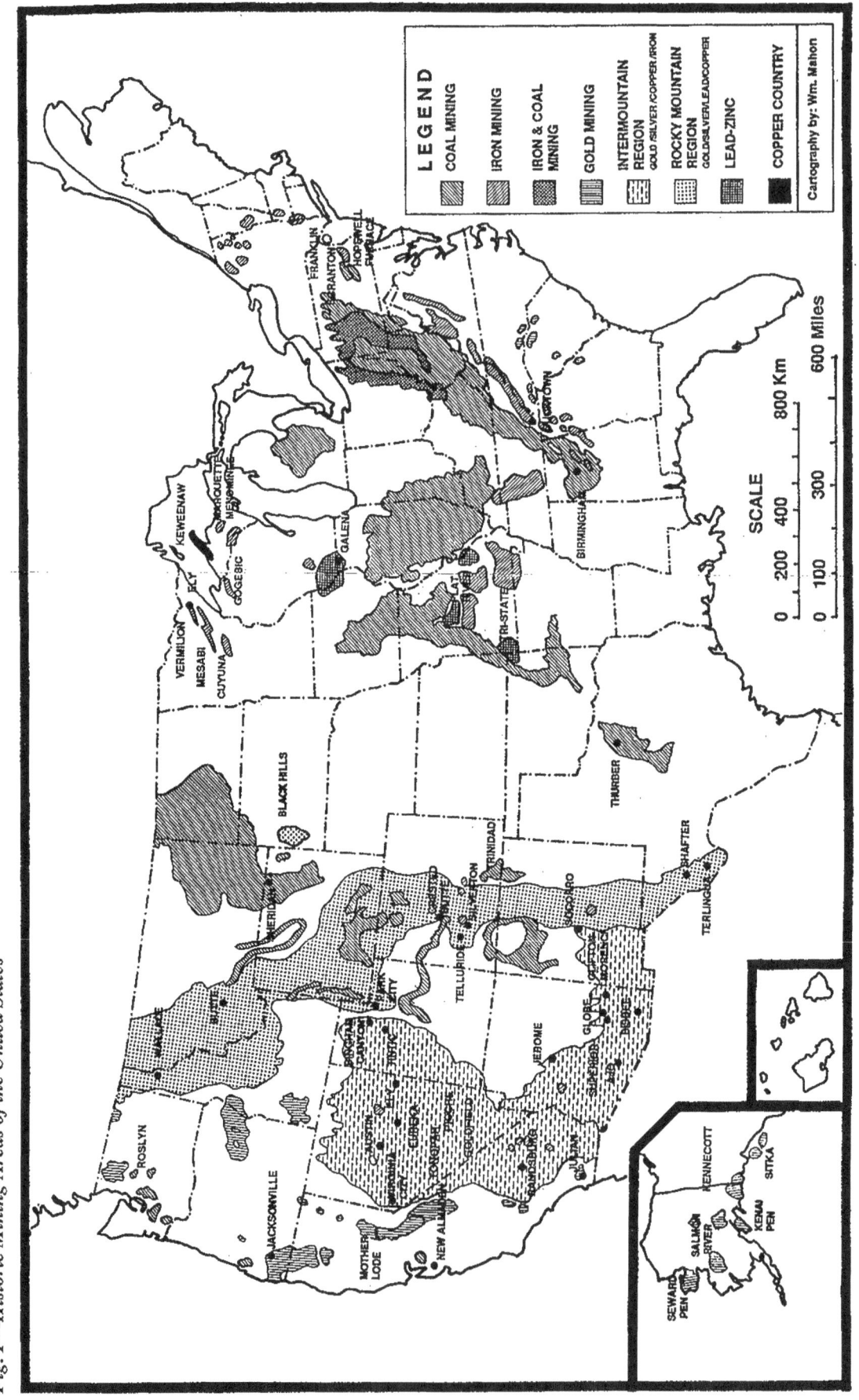

16

mechanically or chemically treated to extract valuable metals. This is the concentration stage of the process. In order to effectively treat the ore, three things must be considered: 1) the major minerals present, 2) relative abundance of minerals, and 3) grain size of various minerals.[7] Then a concentration process is chosen, dependent upon the ore type to be milled.

Crushing

Mineral ore has to be crushed to recover valuable metals. Comminution is the crushing or pulverization of ore into minute particles by mechanical means.[8] This finely-ground material allows recovery of valuable metals. Crushing can be accomplished through a variety of machines. The simplest form of crushing began with arrastras. Arrastras were simple apparatuses with a rock floor and a heavy stone on a central pivot powered by water or a horse. Ore dumped onto the floor was pulverized as the heavy stone passed over it.[9] The arrastra was superseded by the Chilean and Huntington mills that work on the same general principal.

Crushers

Nineteenth-century mechanical crushers included jaw crushers, gyratory crushers, and stamps. Jaw crushers were patented in the United States in 1858. The action of the jaw crusher is similar to the human jaw. There are two plates: one fixed and one moving plate of teeth. The jaw crusher was fed at the top (or mouth) and discharged through the throat. It was an intermittent action. The operator fed the ore into the machine, stopped to collect the crushed ore at the bottom of the machine and dump the crushed ore into bins. Eventually, crushers were placed above bins so that the crushed rock dropped directly into the bin. Jaw crushers were generally considered the primary crushing stage. Gyratory crushers evolved from the concept of the jaw crusher to offer a continuous crushing action. They have remained standard in the crushing phase of milling today.

Stamp mills

Widely used in the nineteenth century, the stamp mill worked on the same principle as a mortar and pestle. Large weights attached to wooden or metal arms connected on a cam shaft were dropped from their highest point onto ore deposited into a trough, often concrete. The stamp could be a dry stamp or wet stamp. The dry stamp functioned as it sounds without water. The crushed ore was filtered through a system of screens. The wet stamp used water to direct a pulp (ore and water) through the trough. As

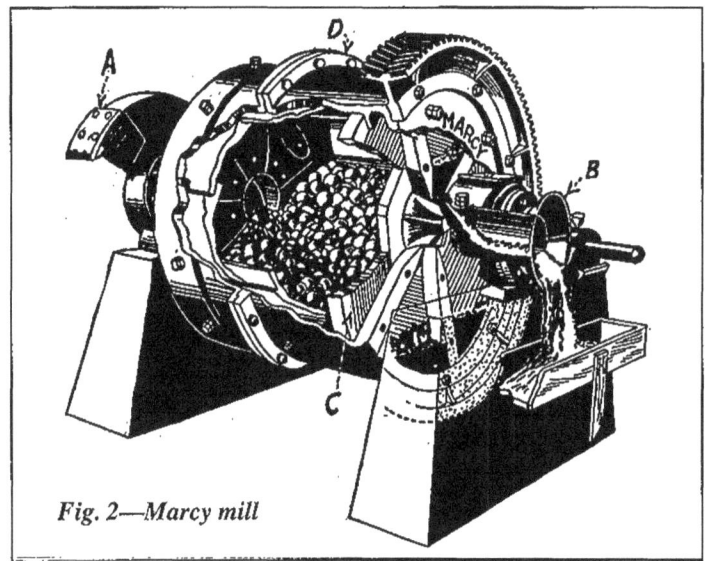

Fig. 2—Marcy mill

batteries were retired with the development of rod, ball, and tube mills, which ground the ore to a fine, sandy to powder state needed for the flotation process. Grinding mills were large, rotating cylinders with steel balls or steel rods crushing the ore inside (Fig. 2). Improved crushing and grinding systems, as the development of flotation, introduced a new trend in mineral milling.[11] The importance of the crushing stage in ore processing was crucial in classification and concentration.

the stamp crushed the ore, the pump and water created a pulsing action with water and lighter waste propelled out of the trough and over amalgamation plates or other collecting devices. Stamps were often used with amalgamation or as an intermediary crusher before a gyratory crushing machine. Stamps and amalgamation were two-step processes used in gold recovery still in use at the turn-of-the-century, although stamps were being replaced in the twentieth century.[10]

Twentieth-century crushing

Experimentation by inventors, usually mining and milling men in the industry, led to the development of improved crushing equipment in the ore process. Without these improvements, flotation would not have been possible. Stamp

Another integral step in the ore process was the classifier. The grinding process was linked to a classifier that sorted the ore by size. If the ore was not crushed or ground finely enough, it was diverted for further grinding before it reached the concentration stage.

Concentration Methods

Concentration is the separation and accumulation of valuable minerals into a concentrated form and is a stage within ore processing.[12] Mineral concentration increases the value of ore by recovering the most valuable minerals in the highest grade possible or purest form of concentrate. Various methods of concentration have evolved over time. There are two major methods to extract valuable metals: pyrometallurgy (smelting) and hydrometallurgy (milling).[13] During smelting the ores are heated until they are molten, with the components separating at their melting points.[14] Milling can be achieved by several methods, with the most common methods of concentration, including amalgamation, concentrators, chlorination, lixiviation/leaching, and flotation. For purposes of this report, we will be dealing with hydrometallurgy, milling using a water bath.

The mill man decides what concentration method should be used for a particular ore body by considering the mineralogy of the ores. There are several conditions to consider: facilities, transportation, cost of smelting, cost of concentration method, amount of recovery, value of concentrate, grade of crude ore, and the contract between the seller and the buyer. Developments in concentration technology have led to the obsolescence of some methods, but the criteria remains integral to the decision process.[15] Flotation will be discussed in more detail in the next section.

After concentration, waste was left where it was extracted.

Gravity Concentration

The most rudimentary form of metal recovery was gravity concentration. These concentrators used water and agitation to sort the heavier metals, which settled to the bottom, from the waste. The most common mechanical concentrators were jigs (the most widely applied gravity concentrating device),[16] vanners, and the Wilfley table. The jig remains of importance in twentieth-century coal cleaning and processing of iron ores. Constructed in 1895, the Wilfley table was a moving, slanted table with riffles on its surface that separated the ore pulp according to the weight of the metal. The pulp was washed over the table, with several minerals collected in the riffles. The waste carried in the water flowed off the bottom of the table into a trough.

The Wilfley table created a technological revolution when it separated minerals. Gravity concentration was a simple and inexpensive process. The biggest problem was that concentrates collected by gravity concentration methods were not pure, but a combination of metals. The Wilfley table and flotation allowed for a higher percentage of mineral recovery and a higher grade of concentrate than earlier methods.[17] The Wilfley table and its principal is still in use in milling today.

Amalgamation

Small particles of free (loose from other elements of ore) gold and silver were collected through amalgamation in the nineteenth century. Amalgamation was the separation of metals by their propensity to adhere to mercury, which formed an amalgam. A crushed ore and water mixture (or pulp) passed over a copper plate covered with a thin film of mercury. These plates were scraped at regular intervals from one to thirty days. The scrapings were placed in a settling pan filled with water. The amalgam (mercury adhered to the metal) settled to the bottom of the pan, while the waste was washed away. The amalgam was then heated, the mercury vaporized, and a base metal product was left behind.[18] Amalgamation was superseded by cyanide leaching beginning in the last decade of the nineteenth century.

Leaching

Leaching was basically the removal of metals in a solution of one or more soluble minerals by percolating liquids passing through the ground ore.[19] Two common leaching methods were chlorination and cyanidation. Cyanidation was the process of

extracting complex ores, particularly gold and silver, by treating crushed ore with a diluted cyanide solution to leach (dissolve) valuable metals from the waste.[20] In 1891, the first cyanidation mills in the United States were built. The mills' flow-sheets (series of operations within the plant) consisted of the following steps: crushing/grinding of ore, concentration of metals from waste (gravity concentration), leaching tanks, filtration, precipitation and recovery. Cyanidation mills offered higher percentages of metal recovery than earlier amalgamation mills. Cyanidation is still in use in the United States today and is found principally in heap leaching (piles of ore with cyanide solution applied to it). Chlorination (developed in 1858) used the general principle of cyanidation, except a chlorine solution was used in the leaching process. It was superseded by the far more efficient cyanidation.

Numerous concentration techniques have been experimented with over time. Many became obsolete; others were in limited use around the world. Leaching has had a resurgence in precious metal mining markets today. Electrostatic and magnetic separators found their use in the recovery of minerals that have magnetic tendencies. A mill survey conducted in 1864 found three types of plants, those milling gold, native copper, and lead. Those mills used free milling, earlier crushing methods and smelting. A second survey in 1895 found mills processing gold and silver, native copper, galena/blende, complex copper-lead-zinc, galena, blende (spholerite), phosphate, sulphide copper, magnetite, oxide zinc, limonite and pyrite. Those mills used gravity concentration and smelting to separate minerals. By 1945, 24 varieties of metallic ores and 19 non-metallic were being processed. Flotation was the key to increased mineral production. Flotation was a successful process in separating more types of minerals than ever possible before.[21]

Endnotes

1. Donald L. Hardesty, The Archaeology of Mining and Miners: A View from the Silver State, Special Publication Series, no. 6, (Ann Arbor, Michigan: The Society for Historical Archaeology, 1988), 18.

2. Richard O. Burt and Chris Mills, Gravity Concentration Techniques (Amsterdam, Oxford, New York: Elsevier, 1984), 5.

3. Richard V. Francaviglia, Hard Places: Reading the Landscape of America's Historic Mining Districts (Iowa City, IA: University of Iowa Press, 1991), 5-9; and Richard Burt and Chris Mills, Gravity Concentration Techniques, 5-7.

4. Ibid, Introduction.

5. Paul Thrush, A Dictionary of Mining, Mineral, and Related Terms (Washington: U. S. Dept. of Interior, Bureau of Mines, 1968).

6. Ibid., 773.

7. Burt and Mills, Gravity Concentration Techniques, 5.

8. Hardesty, The Archaeology of Mining and Miners, 38 and Nathaniel Arbiter, ed., Milling Methods in the Americas (New York: Gordon and Breach Science Publishers, 1964), 7.

9. Hardesty, The Archaeology of Mining and Miners, 39.

10. Robert Richards, Textbook of Ore Dressing (New York: McGraw-Hill Book Co., Inc., 1940), 38-40; Hardesty, The Archaeology of Mining and Miners, 39-41; and A. M. Gaudin, Principles of Mineral Dressing (New York: McGraw-Hill Book Co., Inc., 1967), 43-46.

11. Pierre R. Hines, "Before Flotation," in Froth Flotation 50th Anniversary Volume, ed. D. W. Fuersteanu (New York: American Institute of Mining, Metallurgical and Petroleum Engineering Inc., 1962), 9.

12. Thrush, A Dictionary of Mining, 246.

13. Lysa Wegman-French, "The History of The Holden-Marolt Site in Aspen, Colorado: The Holden Lixiviation Works, Farming and Ranching, and the Marolt Ranch 1879-1986" (Aspen, CO: Aspen Historical Society, Oct 1990), 13.

14. Ibid.

15. Flotation Fundamentals and Mining Chemicals (Midland, MI: Dow Chemical Company, 1968), 91 and Gaudin, Principles of Mineral Dressing, 1.

16. Burt and Mills, Gravity Concentration Technology, 184.

17. A. M. Gaudin, Flotation (New York: McGraw-Hill Book Co., Inc., 1932), 531; Burt and Mills, Gravity Concentration, 3; and Hines, "Before Flotation," 5.

18. Hardesty, The Archaeology of Mining and Miners, 46-47; Thrush, Dictionary of Mining, 32 and Richards, Textbook of Ore Dressing, 52-53.

19. Thrush, A Dictionary of Mining, 630.

20. Hardesty, The Archaeology of Mining and Miners, 51 and Thrush, Dictionary of Mining, 296.

21. Nathaniel Arbiter, ed., Milling Methods in the Americas (New York and London: Gordon and Breach Science Publishers, 1964), 8.

Chapter 3 – History of Flotation

Modern engineers consider flotation the beginning of modern mining. Initially, man's mining endeavors began with the desire to locate precious metals of gold and silver. A twentieth-century American industrial age initiated even more experimentation with metals in manufacturing. Before the end of the nineteenth century, the mining industry was entering a crisis period. Increased levels of mining in the United States had greatly reduced high-grade ore bodies (ore bodies with a high percentage of a particular metal). Simple ores, such as copper and iron, had been mined for eons of time. Remaining mineral resources were complex ores containing smaller quantities of valuable metals (low-grade). The metals in low-grade ore bodies had long been recognized; but the problem was how to economically separate the metals into quantities large enough to be profitable.

The development of the cyanide process at the end of the nineteenth century allowed low-grade gold ores to be treated effectively. However, the world had come to rely on many minerals other than gold. In order to separate base metals in low-grade ore, a new concentration method had to be found. Prior concentration methods (gravitation, leaching, and smelting) provided two products, concentrate and waste. As a result, these methods were ineffective in separating metals in complex mineral bodies. The mill man's objective was to separate different, valuable industrial minerals from each other into several concentrates. Experimentation proceeded at a furious pace looking for a new concentration method to treat low-grade ore.

The ensuing development of flotation allowed for the most efficient means of mineral recovery to date. The mining and milling industry had found a way to exploit complex minerals that were impossible to treat by earlier gravity methods. Without flotation, metals (such as copper, lead and zinc) would have become increasingly costly and difficult to produce causing a direct impact on global economy.[1] In addition, flotation made new resources commercially available on the minerals market. It, also, can be combined with other processes in production, both within and outside the mineral industry. Modern engineers and inventors assert that the evolution of a flotation process in the early decades of the twentieth century was the most important development of the century in the recovery of metals.[2]

Flotation

Flotation is a method for concentrating valuable metals from finely ground ores in a water bath, with either the ore pulp or water chemically altered with reagents and frothers to encourage separation of minerals. The term flotation has been loosely used for all concentration processes in which heavier mineral particles have been separated from lighter waste particles in water by "floating" the mineral away from waste. Today the term is generally used to describe froth flotation, but it is necessary to understand that flotation evolved through three principal stages of development. (These stages will be discussed in more detail later.) The process of flotation involves: 1) the reduction (crushing) of the ore to a size that will free the minerals desired; 2) the addition of water, reagents and frothers, which encourages the minerals to adhere to air bubbles; 3) the creation of a rising current of air bubbles in the ore pulp; 4) the formation of a mineral-laden froth on the water surface; 5) skimming or floating off the mineral-laden froth; and 6) the drying of the resultant concentrate for shipment to the smelter.[3]

Flotation relies on very different principles than earlier mill devices that used differences in specific gravity. Flotation separates

materials by taking advantage of the surface tension of liquids and the ability of minerals to attach to air bubbles in liquids.[4] Another innovation of the flotation process is its ability to chemically alter the pulp (in the water bath or ore pulp itself) allowing for an efficient separation of numerous minerals from waste without increasing the tonnage or grade of ore.[5] Flotation is used in processing non-ferrous metallic ores, ores not containing iron. With minor adjustments in the chemicals added to the water, the process can be repeated as often as necessary to recover as many different types of metals or minerals from the ore pulp. Flotation, as developed in the twentieth century, has now become the most widely used process for extracting minerals from metallic ores.[6]

In the following sections, I will discuss: the evolution of flotation in the metals industry; the inventors; when general industrial use of flotation began; what flotation circuits are; and its current use in the milling industry.

Development of Flotation

Nineteenth-century mill men acknowledged that current milling practices, i.e. crushing, gravity concentration and smelting, were not efficiently and economically recovering all the mineral wealth from extracted ores. They anticipated that high grade ores would soon be depleted. Earlier milling methods recovered one mineral or a mineral compound, e.g. lead, zinc, or lead-zinc. The dilemma was how to recover many separate minerals from low-grade ore into a high-grade concentrate.

Experimentation flotation in complex ore concentration began in England in 1860 with William Haynes. Haynes found, when mixing powdered ores, oil, and water, some minerals had a tendency to attach to certain oils.[7] By the end of the century, many others experimented with such variables as: 1) additives, such as oils, acids, or salts; 2) agitation; and 3) heat. Several individuals patented their findings in the late 1800s. Carrie Everson's patents (1886 and 1891) included a two-step process thoroughly mixing pulverized ore, oil, and an acid or salt together into a pulp, then agitating the pulp (crushed ore and oil) on an irregular work surface. In earlier experiments, Everson had used a washboard as a work surface. The metals rose to the top of the work surface and the waste settled to the recessed areas of the surface. In her patent, Everson listed the minerals she successfully recovered.[8] The Wilfley table, developed in 1896, works on the same principal as Everson's irregular work surface. By 1894, George Robson and Samuel Crowder working at a mine in Wales developed a mixture balance with oil, which resulted in

the oil lifting mineral particles to the surface. Their process was the forerunner of the Elmore bulk oil process.

Bulk Oil Flotation

The early development of bulk oil flotation was attributed to the Elmore brothers, Francis and Alexander, of England in 1898. The two men patented various modifications of a process involving equal parts of ore and oil in a drum. The contents were poured into a water trough, the concentrate rose to the surface, and the scum skimmed off the top.[9] Francis went to work for the English-based company, Minerals Separation Incorporated, where he continued his efforts to perfect the flotation process. In 1904, Elmore and Minerals Separation patented the introduction of bubbles, which subsequently led to another patent based upon those same mineral-clad bubbles rising to the surface under a vacuum.

Elmore's experimental plant led to the construction of the first commercial flotation mill in Broken Hill, Australia in 1905. Soon after, the Elmore process could be found in mills in Australia, South Africa, Canada, England and Wales.[10] Although all the Elmore-process mills eventually failed, due to excessive amounts of oil that were used,[11] the process was found in limited use in Australia as late as 1938.

Other experiments in bulk-oil flotation led to the use of gas bubbles and an altered chemical solution. Italy's Alcide Froment laid claim to the first patent (1902) to mention the use of gas for increasing buoyancy of metallic particles. A. E. Cattermole's patent in England in 1903 used less oil, introduced soap and alkali into the pulp stream, and agitated the pulp in a water bath through a continuous stream of gas bubbles. Although his process was a commercial failure, Cattermole's process is considered the forerunner to modern flotation methods used today.[12]

Skin Flotation

Another type of flotation used to separate minerals from waste, without the use of oils, was skin flotation. In June 1885, Hezekiah Bradford patented a process for separating sulfide ore based upon the principle of surface-tension on water.[13] (Sulfide ores can contain copper, lead, zinc, iron, molybdenum, cobalt, nickel, and arsenic.) The process, commonly called skin flotation, stipulated that dry ore be sprinkled, without agitation, onto the surface of a body of water. The heavier materials sank to the bottom, while the surface tension caused the lighter metals or minerals to float on the surface. Subsequent developments by de Bavay and Elmore in 1905 made this process commercially significant. Their

process used a small amount of oil (fraction of 1 per cent) mixed with one ton of mineral pulp. The mixture was then poured onto a water bath and agitated to encourage frothing. The Macquisten tube (named for its inventor) mixed the ground ore with the oil and gently pushed the pulp onto the surface of the water. The tubes were first commercially installed in 1906 in the Adelaide Mill at Golconda, Nevada. In 1911, the Morning Mine near Mullen, Idaho used the process on lead-zinc ores in its mill. As the process was extremely delicate, its use was short-lived. The Idaho mill closed by 1920.[14]

Froth Flotation

The first successful commercial mill using froth flotation was built in 1905 at Broken Hill, Australia, today recognized as the "home of flotation." Independent experimentation, based upon earlier patents, at Mineral Separation's Broken Hill mines and mills resulted in the froth flotation process and the Minerals Separation Machine. In 1905, Mineral Separation engineers, E. L. Sulman, H. F. K. Picard, and John Ballot, filed a patent for their process that separated froth flotation experiments from skin and bulk-oil flotation. Their froth flotation process (based on Froment's and Cattermole's earlier patents) used less oil in the ore pulp and added agitation by a rising stream of air bubbles. In another Minerals Separation mill, a new innovation was the use of a Minerals Separation Standard Machine, which was a spitz box invented by Theodore J. Hoover (brother of American President Herbert Hoover). The pulp passed over the spitz box causing the mineral-laden bubbles to rise to the surface. This in turn allowed the collection of mineral-laden froth from the water. The Minerals Separation Standard Machine consisted of a series of cells divided into agitation and frothing compartments. Although it was eventually superseded as improvements were made to the machine, Hoover's Separation Machine significantly advanced mechanization of the milling industry.

New technology in crushing and separation of ground ore further advanced the development of flotation. While mechanical classifiers, separation machines and the Wilfley table assist in the separation phase of the flotation process, selective flotation requires finely ground ore (sand or dust) to allow metals to be recovered on the froth of a water bath. Without improved grinding treatment in ball and rod mills, flotation would not have been possible.

Inventors Sulman, Picard and Ballot continued to experiment with soluble frothing agents (commonly referred to as reagents) and patented several combinations used at the Broken Hill mills. Aware that many minerals react differently to reagents in the water bath, experimentation continued to create formulas necessary for the separation of numerous minerals. The isolation of desired metals would depend on the delicate balance of acid and alkaline agents in the flotation process. However, Minerals Separation's first froth flotation plant at Broken Hill was a success.

Flotation Circuits

Flotation plants operate through a series of flotation circuits. Finely crushed and ground ore is introduced into a water bath in the flotation circuit. In order to achieve separation, an optimum point has to be met. There are several variables that need to be considered: particle size, reagent additions, pulp density, flotation time, temperature of the pulp, type of circuit, and water. Next the ore itself must be considered: uniformity of the ore, settling and filtration data, corrosion and erosion, and finally, the mineralogy of the ore.[15] Once all these variables are tested in pilot plants, the concentrator flotation circuits are designed.

Generally, there are two types of separation circuits, a simple or a complex circuit. Simple separation circuits are generally described as one-product separations. A single

27

Fig. 3—Flotation division, Timber Butte mill

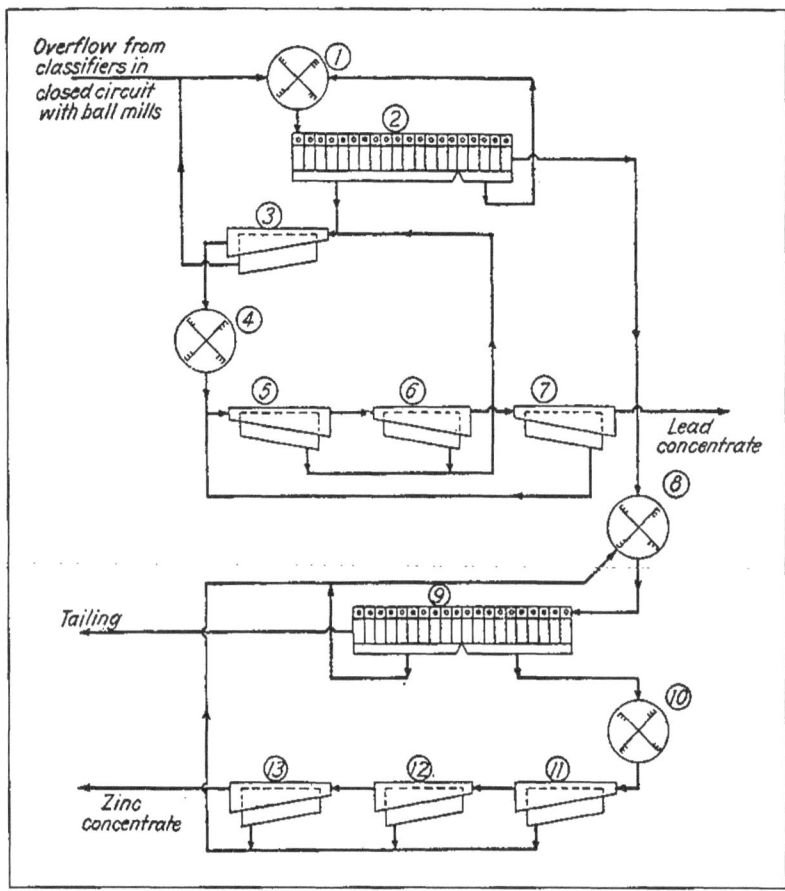

①	classifier "surge tank"
②	20-cell minerals separation flotation machine
③	callow cell (flotation machine)
④	conditioning tank
⑤⑥⑦	callow cells "roughing", "cleaning" & "recleaning"
⑧	conditioning tank - reagents added
⑨	minerals separation flotation machine
⑩	surge tank
⑪⑫⑬	callow cells

circuit separates the ore pulp into a concentrated product and tailings. In order to process more than one concentrate, the one circuit mill and its workings are periodically adjusted to recycle the ore pulp through the system. After chemically altering the bath or pulp, the process can be repeated as many times as necessary to float out all desired metals. (For a description of a single circuit refer to Figure 3.)

A complex flotation circuit separates several concentrates from the ore pulp. Valuable metals are floated together to form a bulk product. The bulk product continues its course through a series of flotation cells where metals are "selectively" separated into concentrates. The waste is sent off to the tailings pond. The complex or selective flotation circuits are essentially several cells in a series. The first circuit of cells may separate lead and copper, the second zinc, and the third an iron reject. Yet, the flow of the pulp through the flotation circuit is continuous.

Additional operations in the mill include roasting, steaming, thickening, filtering, leaching, and

28

regrinding of the ore pulp in the flotation circuit.[16] For an example of a complex set of circuits, refer to Appendix A, for the flow sheet of the Shenandoah-Dives Mill in Silverton, Colorado. Plants designed similarly to the Shenandoah-Dives did not necessarily produce the same concentrate, even on the same ore. Adjustments were made during the process to allow concentration of different grades of ore. Testing of ore was always an on-going process. It is important to note that a circuit can separate zinc one day and be adjusted to separate copper the next day, depending on the needs of the mill and the mineralogy of the ore it was processing. Flexibility of the concentrator (mill) to simultaneously treat various grades of ore improved the mill's ability to stay viable in fluctuating markets.

Equipment found in a complex circuit may include: roughers, flotation cells (in many sizes), launders, thickener tanks, regrinders, reagent or frother machines, tubes for roasting or heating pulps, pumps, samplers, settling boxes, filters, and concentrate bins.

Commercial Milling

In the early decades of the twentieth century, new patent applications for innovative milling processes were filed almost yearly, in some cases monthly. In the pursuing years, litigation suits, generally contesting whose idea came first or concerning wording in patents, tied up the courts for years. Nevertheless, experimentation in flotation continued to take place in small plants around the world. Sulman, Picard and Ballot perfected their froth flotation process in a pilot plant in Broken Hill, Australia. An American engineer who traveled to Australia in 1911 to study the Broken Hill mills found four different milling methods in use.[17] They were: froth flotation, wet-film process, agitation-froth flotation and the Elmore vacuum.[18] Eventually, only froth flotation remained as the clear winner.

In August 1911, James Hyde, familiar with the English-based Minerals Separation's work in Australia,[19] experimented with froth flotation in the Basin Reduction Co. Mill in Basin, Montana. The Basin Mill was leased by the Butte and Superior Copper Company for Hyde's experiments. When the lease expired in 1912, Butte and Superior designed and built a mill at the Black Rock mine in Butte, Montana. This mill was recognized as the first commercial froth-flotation plant in the United States. The facility used gravity concentration in the first three stages and froth flotation in its final stage.[20]

By 1914, there were 42 American mining companies using froth flotation in their mills.[21] Skin and bulk-oil flotation soon became obsolete. The first mill to use only the flotation process in its concentration process was the Engels Mine and Mill in Plumas County, California in 1914.[22] J. M. Callow at Miami Copper Company in Arizona developed and marketed his own version of pneumatic flotation cells, similar to the Minerals Separation Standard machine.[23] Mills continued to experiment with reagent use in froth flotation process on a variety of ores. Chemical agents, acid or alkaline, in the flotation process affected metal recovery in various ore bodies. The Sunnyside Mill at Eureka, San Juan County, Colorado was credited with perfecting the flotation of two minerals (lead-zinc) from a complex ore in 1918.[24] C. H. Keller introduced xanthates (water-soluble froth collectors) into water baths in flotation circuits in 1925.[25] In 1932, G. K. Williams introduced the first continuous refining process.[26] However, by the end of the 1920s, the fundamentals of flotation had been established.

The development of flotation had a major impact on the mining world. In concentration, flotation took over much of the field from former methods of gravity concentration, amalgamation and leaching. In his 1932 text, Flotation, A. M. Gaudin attributed major changes in mining and milling metals due to the development of flotation

concentration. Flotation increased recovery and grade of concentrates and decreased the cost of metal production and the price of metals in the market place. More importantly, Gaudin asserted flotation increased the use of metals in American industries.[27]

By the 1930s, froth flotation provided new commercial resources. Low-grade ore and materials once considered waste were processed profitably. Arrival of poly-metallic mills allowed the separation of more than one metal in a series of flotation circuits. Single circuit plants continued to be useful in many regions. New poly-metallic mills were larger in size, generally processing 750 to 2,000 tons per day, to accomodate multiple circuits and ore bins. Strong companies in the industry made a move to consolidate mineral patents. Mills and their equipment also expanded in size and processing capability. In the 1940s, mining companies also increased their use of open pit mines, which greatly increased the amount of ore extracted. As a result of increased ore extraction, the size of mills grew exponentially to accommodate ore delivered to the mills. The new mills had multiple buildings in complexes that covered acres of land.

Mill expansion resulted in the replacement of obsolete or

exhausted operations. By the 1960s, larger electrical capacity generators were designed, which allowed for variable speed drives in mills. Even larger sizes of mill equipment could be built and operated on this increased horse power. New equipment allowed for the addition of automatic controls and instrumentation. Where small mills were not capable of developing adequately to keep up with large mine operations, new plants were constructed on old mill sites.

In the year 1960, 202 flotation plants were operating in the United States. These mills processed: metallic minerals (96), nonmetallic minerals (67), bituminous coal (26), anthracite (5), waste paper (3), and miscellaneous materials (5). Of the metallic-mineral mills, 30% processed copper, lead and zinc. An additional 26% processed lead-zinc.[28] In addition, flotation allowed other industries to alter the process to fit their particular needs, whether it was non-metallic minerals, coal or paper industries.[29]

Conclusion

Flotation had a profound effect on the mines and minerals industry, as well as industry as a whole, by providing all the needed metals at a lower price than otherwise would have been possible. Principally, it shifted the focus of the mining industry from extraction of high-grade ores to one featuring

complex and low-grade ores. Flotation essentially reorganized the industry. It allowed non-selective mining and selective milling practices. Vast quantities of metals were extracted from underground mines, while the most valuable metals were separated in the mill on the surface. The mine was able to extract ores that had minimal quantities of valuable metals, often difficult to extract, locked in tons of waste rock. In the flotation process, the millman was able to selectively separate metals that were profitable or desirable from the waste. Flotation, also, increased the recovery and grade of concentrates without increasing the tonnage or the grade of the ore. In 1962, mining engineers Charles Merrill and James Pennington wrote an essay on new resources available in U.S. flotation mills, which was included in Froth Floation, 50th Anniversary. They found that flotation increased production with 24 metallic and 19 non-metallic ores since the commercial success of flotation in 1905. Many modern engineers proclaim flotation as the most important twentieth-century development in the recovery of metals.

Endnotes

1. Jeremy Mouat, "The Development of the Flotation Process: Technological Change and the Genesis of Modern Mining, 1898-1911," Australian Economic History Review 36, no. 1 (March 1996): 4.

2. "The Trend of Flotation," Colorado School of Mines Magazine (January 1930): 20; Mouat, "Development of Flotation," 4; author's phone interview with Richard Graeme on 10 July 1997 concerning developments in mining in the twentieth century; and Charles White Merrill and James W. Pennington, "The Magnitude and Significance of Flotation in the Mineral Industries of the United States," in Froth Flotation 50th Anniversary (New York: American Institutes of Mining, Metallurgical and Petroleum Engineering Inc., 1962), 56-57.

3. Flotation Fundamentals and Mining Chemicals (Midland, MI: Dow Chemical Co., 1968), 8.

4. Mouat, "Development of Flotation," 6 and Flotation Fundamentals, 11.   .

5. Thomas Rickard, Concentration by Flotation (New York: John Wiley and Sons, Inc., 1921), 408.

6. E. H. Crabtree and J. D. Vincent, "Historical Outline of Major Flotation Developments," in Froth Flotation 50th Anniversary Volume, ed. D. W. Fuerstenau (New York: The American Institute of Mining, Metallurgical, and Petroleum Engineers, Inc., 1962), 39.

7. Robert Richards, Textbook of Ore Dressing (New York & London: McGraw-Hill Book Co., Inc., 1932), 233 and Crabtree and Vincent, "Historical Outline," 39.

8. Rickard, a leading engineer in twentieth-century mining, disputed Carrie Everson's stature as an innovator in the milling world. He did not disclaim her findings, but was distressed that the findings were attributed to Carrie Everson. Rickard wrote several articles discussing the topic including an article called, "Everson Myth" in the January 15, 1916 issue of the Mining and Scientific Press. Rickard was editor of the journal.

9. Crabtree and Vincent, Froth Flotation, 40.

10. Rickard, Concentration, 8.

11. C. Terry Durell, "Universal Flotation Theory," Colorado School of Mines Magazine 6 (February 1916): 27.

12. Ibid, 28; Crabtree and Vincent, Froth Flotation, 42.

13. Theodore J. Hoover, Concentrating Ores by Flotation (San Francisco: Mining and Scientific Press, 1914), 20.

14. Crabtree and Vincent, Froth Flotation, 40.

15. Adrian C. Dorenfeld, "Flotation Circuit Design," in Froth Flotation 50th Anniversary (New York: American Institute of Mining, Metallurgical and Petroleum Engineering Inc., 1962), 365.

16. Ibid, 370-1.

17. Durell, "Universal Flotation Theory," 29-30.

18. Ibid, 29-30; Crabtree and Vincent, Froth Flotation, 42-43.

19. For details on Hyde's familiarity with Minerals Separation's process refer to Pierre Hines and J. D. Vincent's essay, "The Early Days of Froth Flotation," in Froth Flotation 50th Anniversary Volume, ed. D. W. Fuerstenau (New York: American Institute of Mining, Metallurgical and Petroleum Engineering Inc., 1962).

20. Burt and Mills, Gravity Concentration, 17 and Hines and Vincent, "Early Days of Froth Flotation," 12-14.

21. Hines and Vincent, "Early Days of Froth Flotation," 29, citing G. A. Roush, The Mineral Industry, 1914 (McGraw-Hill Book Co., New York, 1915).

22. Hines and Vincent, "Early Days of Froth Flotation," 28.

23. Richards, Textbook of Ore Dressing, 234.

24. Hines and Vincent, "Early Days of Froth Flotation," citing Editorial, Engineering and Mining Journal 125 (Feb. 4, 1928): 195.

25. Crabtree and Vincent, "Historical Outline," 45.

26. Cedric E. Gregory, A Concise History of Mining (New York: Pergamon Press, 1980), 140.

27. A. M. Gaudin, Flotation (New York: McGraw-Hill Book Co., Inc., 1932), 534.

28. Merrill and Pennington, "The Magnitude and Significance of Flotation," 56-57.

29. Flotation is used in the processing of copper, copper-molybdenum, copper-lead-zinc, copper-zinc-iron-sulfur, gold-silver, lead-zinc, barite, bastnaesite, calcite, garnet, kyanite, and talc. Clay, feldspar, mica, quartz, spodumene, waste paper are processed by flotation. Flotation concentrates fluorspar, glass sand, ilmenite, magnesite, iron oxide, manganese, tungsten, phosphate, anthracite, bituminous coal, chicle, naphthalene and oil.

Chapter 4 – Historic Flotation Milling Properties

This section focuses on property types that may be found in twentieth-century flotation mills. Milling landscapes are reflective of a particular region's terrain and ore body. Sites can consist of interlocking buildings perched on the sides of steep mountain slopes in the Rocky Mountain West to taller, compact mill buildings found in dry desert regions of Arizona and New Mexico. Mills are often found in remote locations of the United States, but increased population has brought them into close proximity to some towns and cities. Butte, Montana is an example of a community (eventually a city) that almost encapsulates the mines and mills that fueled its economy. In Utah, the Midvale Plant was surrounded by the burgeoning population of nearby Salt Lake City. Eventually, the mill was removed due to community concerns of mineral contamination.

In the first decades of the century, most mills (also referred to as concentrators, concentration plants, or reduction plants) were generally located in or near remote mining districts. Twentieth-century milling districts were chosen for their mineral wealth (mine), proximity to a water source

for mine and mill use, and their potential accessibility to transportation (railroad or road systems). A mill built close to the mine and/or smelter kept the cost of transportation of ore and concentrates to a minimum. A labor supply or available property for a community for employees was also of utmost need to the mill. In addition, a large site for disposal of tailings was essential if a mill was to operate for years or even decades.

Mill Building Designs

Many twentieth-century flotation mills were engineer-designed complexes. Engineers designed mills for cost-efficiency. As the National Park Service's publication Guidelines for Identifying, Evaluating, and Registering Historic Mining Properties states, "They [mills] were intricate industrial operations with every component ideally working in harmony to reduce costs, increase production, and maximize profits."[1] Cost efficiency and profit margins were of utmost importance to milling companies and their shareholders. Mill buildings came in a variety of sizes, from the size of a small house (50-100 tons-per-day production) to the modern industrial size covering acres

(60,000 tons-per-day production). Two factors that impacted mill design were terrain (the site where the mill was to be built) and the type of mineral to be processed.

In an essay discussing the trends in mill design up to the 1960s, Norman Weiss and John Cheavens noted wooden structures from the nineteenth century and early twentieth were able to accomodate flow sheet changes with little alterations. By the 1930s, mills were erected with steel and concrete to encompass a fireproof structure. As the milling industry expanded in the 1950s and 60s, old structures were removed and replaced with numerous steel and concrete buildings for crushing, grinding, flotation, roughing, and thickening processes.[2] While this was the trend, smaller operations tended to be frame buildings, with newer buildings often using a combination of frame, steel, and/or concrete.

Basically, mill buildings were built around the machinery, not vice versa. An ore's mineral contents affect the milling process used, whether it be flotation, leaching, or chlorination. Only after the mill process was chosen and a review of the necessary equipment completed was the mill building designed. Allowances were generally made for additional space for new equipment (to replace old or allow for new technology), enlargement of

flotation circuits, or other mill needs. After the foundation was constructed at the mill site, the processing equipment was delivered and installed. A single piece of equipment could weigh several tons with individual parts weighing almost one ton. Then the exterior walls of the building were built around the equipment and foundation.

Natural Features of Mill Sites

Mining engineers designed mill complexes for sloping ground, flat ground, or partly inclined and flat properties:

- In mountainous regions, sloping-ground mills were built as multi-level buildings on the side of the mountain. (See Fig. 4.) The step pattern of the building used gravity to move the ore, which saved power costs. Gravity caused the ore to fall from a crushing machine into a trough that led to a conveyor that carried the ground ore to the mill where it fell into secondary crushers and so on. The gravity flow of the watered ore pulp carried it from classifiers to flotation machines and so on. Thus, there was less need for pumps, elevators and launders.

- Flat-ground mills were generally taller and more compact buildings. (See Fig. 5.) These facilities used more

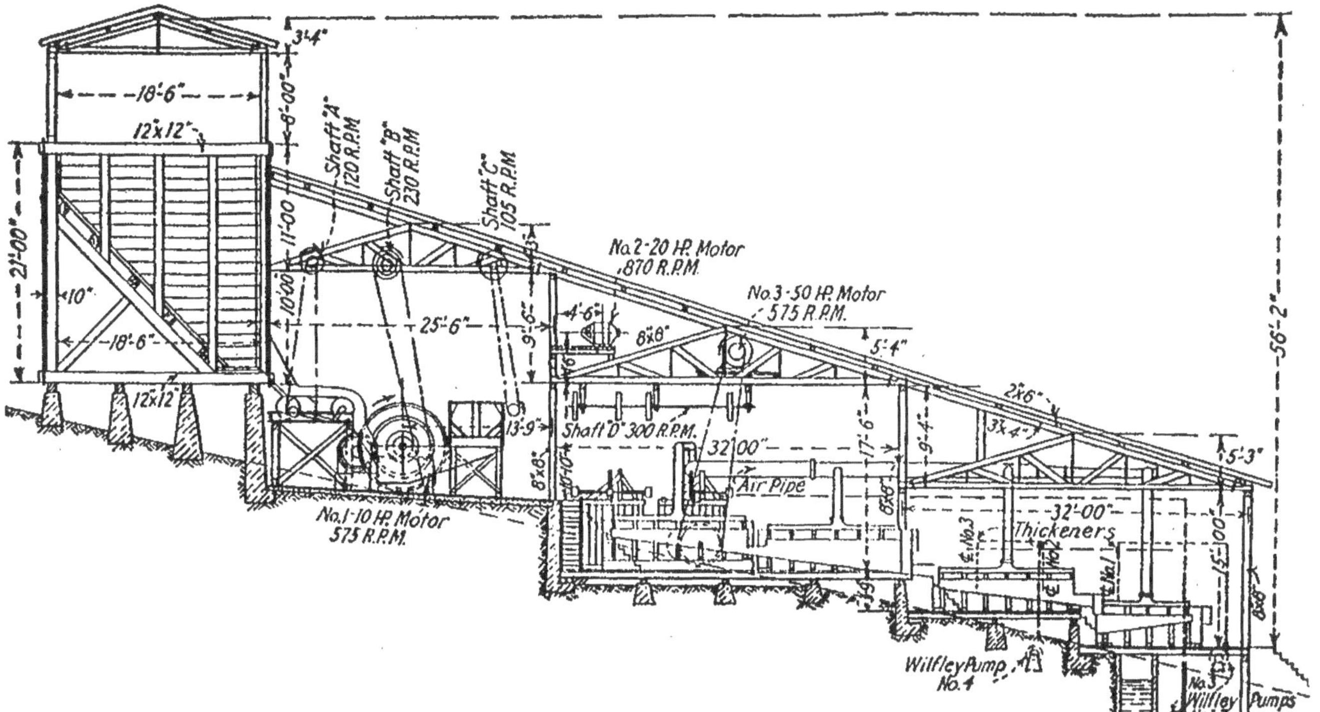

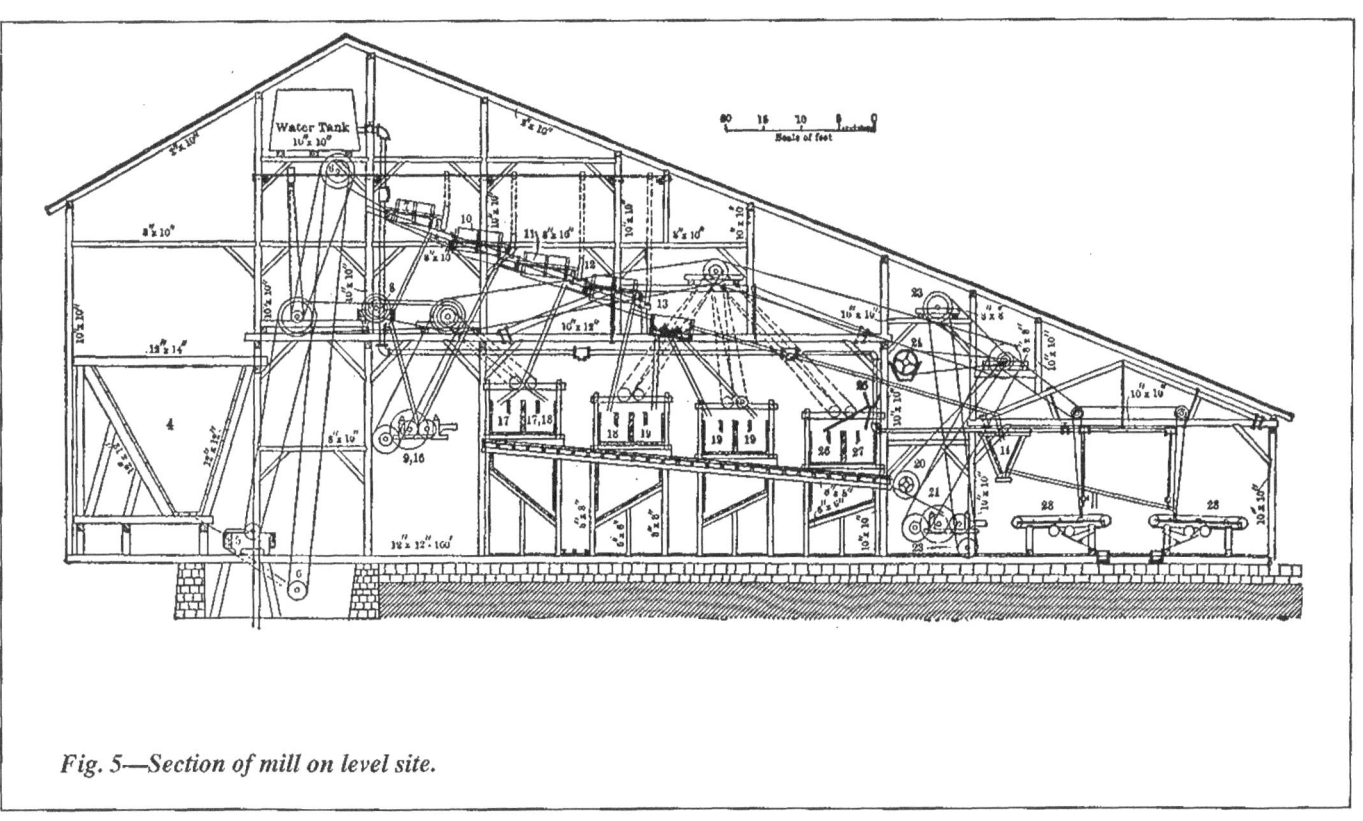

Fig.4—Multi-level design for small lead-zinc plant on a hillside. Typical design for fine-grinding & flotation divisions to handle 100 tons of lead-zinc ore per day.

Fig. 5—Section of mill on level site.

electrical or steam power than their mountainous counterparts to pump the ore through a series of launders and elevators to the various stages in concentration.

- Partly-inclined and flat-ground mills were designed to benefit from both gravity and flotation. The slope was used for the grinding stage and the flat section housed the flotation cells.

Property Types

Milling properties may include several buildings, structures or systems to support mill operations. Some buildings housed more than one system of operations. Generally, mill sites include the following:

- Conveyance to the Mill: aerial tramway, railroad line, truck route

Ore Bucket on Aerial Tramway

- Arrival Terminal

- Ore Bins

- Conveyors: may traverse from bins to crushing plant and from crushing plant to mill.

Arrival Terminal for the Aerial Tramway at Shenandoah-Dives Mill

- Crushing/Grinding Plant: equipment may include jaw crushers, rod or ball mills. Usually the fine grinding plant is located some distance from the primary crushing step, possibly with a storage bin in between.

Coal Bin and Ramp

- Mill Building: designed around the equipment inside (not the other way around), such as classifiers, ore bins, fine crushers, pumps, water pipes, electrical devices, flotation cells, vats, drying cells, thickener tanks, Wilfley tables or similar apparatus, concentrate bins, instrumentation (control panels), reagent distributors, communication network (phone and their lines).

- Power Source: may include electrical transformers, steam power plants, coal bins, pump houses.

- Water Source: pipes or tanks

- Administrative and Operations Office

- Machine Shop/Workshop

- Laboratory/Assay Office

- Warehouses

Electrical Transformer

- Storage Building for chemicals, such as reagents, lime and general storage needs

- Change House building for laborers to shower and change their clothes

- Tailings Ponds

- Conveyance to Smelter: railroad, road system (gravel, dirt, paved)

Historic mill properties can be found perched on the sides of mountains, in the rolling hills of the midwest, and the flat deserts of western states. Over time, many of these abandoned historic mill properties have been altered by several factors, including humans or weather.

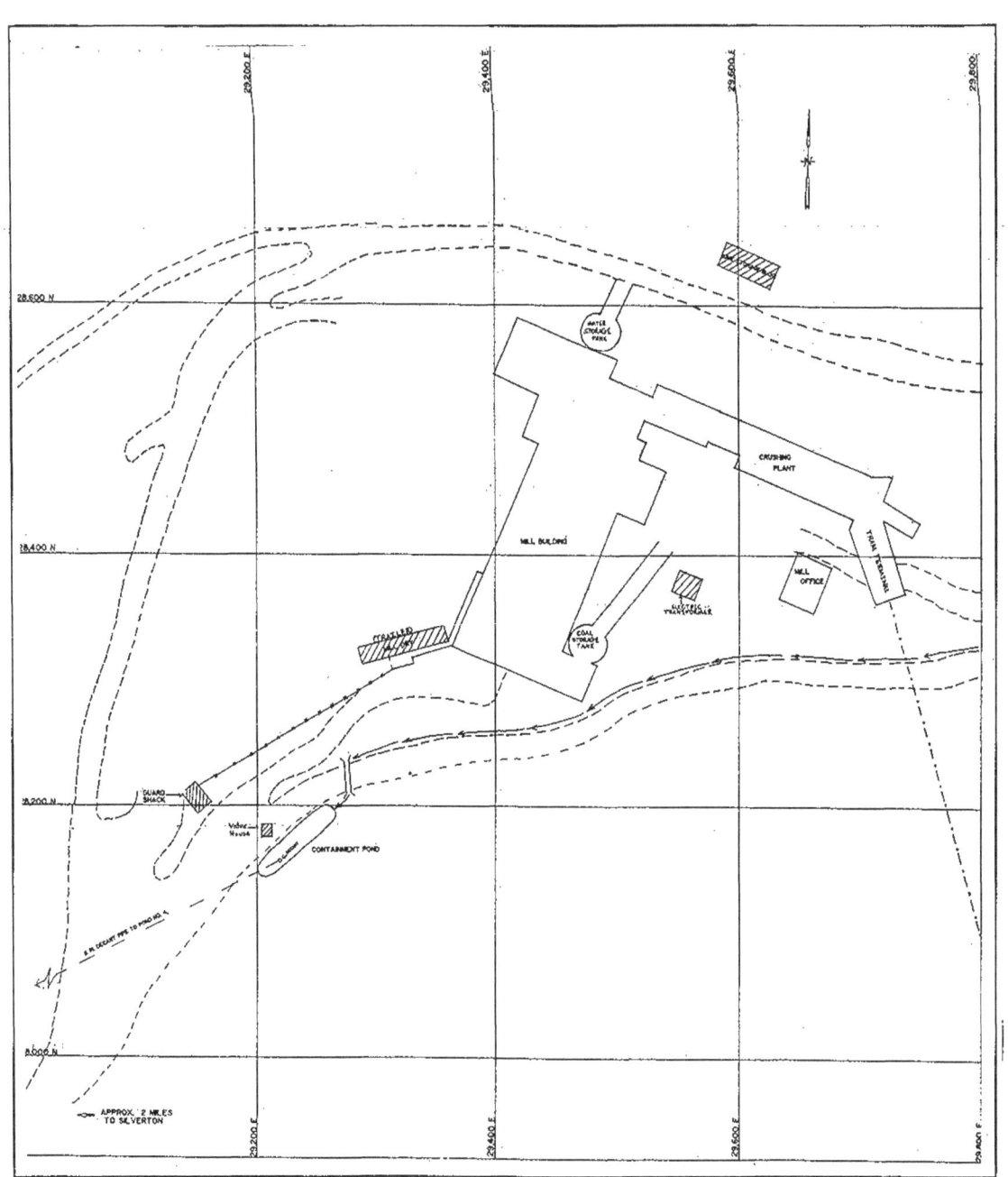

Fig. 6—Shenandoah-Dives Mill (Mayflower) Mill Site Plan

Aerial Tramway Tower and Avalanche Deflector

Adverse Conditions Affecting the Integrity of Historic Mill Properties

Some characteristics that may contribute to or detract from a mill site's physical condition include its accessibility to the intrusive forces of humans or weather. Remote mill sites are not easily reached by highways, thus saving elements of milling history from vandalism. Weather has affected wooden mill properties that deteriorate over time due to heavy snowfall in mountains, rain, and humidity. Many of the wooden structures in mountainous regions have also been lost in avalanches.

Social and natural factors also affect mills. The major ones include:

- Technological advancements and upgrading of facilities have resulted in changes to the original equipment and process flow. In extreme cases, it has led to the bulldozing of earlier concentration plants and mills for the construction of new plants. In many facilities, one can easily detect the overlay of several technologies. As one ore body was depleted or markets fluctuated with demands for various metals, different processing facilities were needed. Mills originally designed for cyanidation might convert to flotation. In other cases, there could be a new cyanide circuit simultaneously running next to a flotation circuit.

- Metal shortages led to scrap drives during the world war years. Consequently, the drives resulted in the destruction and dismantling of mills to obtain valuable metal equipment.

- Property owners have found increasing market demands in foreign countries for out-dated or mothballed American milling and concentration equipment. These sales have resulted in the loss of a large segment of American historic mining and milling equipment and technology.

- Changing political and environmental concerns have led to the destruction of mine and mill sites in industrial cleanup sites, the reclamation of abandoned lands, and the removal of "eyesores" containing deteriorating plants.

Endnotes

1. U.S. Department of Interior, National Park Service, Interagency Resources Division, Guidelines for Identifying, Evaluating, and Registering Historic Mining Properties, by Bruce J. Noble and Robert Spude, National Register Bulletin 42 (Washington, D. C.: Government Printing Office, 1992),13.

2. Norman Weiss and John Cheavens, "Present Trends in Mill Design," Milling Methods in the Americas, ed. Nathaniel Arbiter (New York: Gordon and Breach Science Publishers, 1964),12.

Chapter 5 - Evaluating Historic Mill Properties

Preservation planning includes identifying historic properties and evaluating their significance. Establishing a measure for these properties is complicated due to the relatively few mill sites found in the United States available for this study. Where milling has been continuous, modernization has altered the interior workings of early twentieth-century mills. Where abandonment has occurred, equipment and even buildings have been salvaged. While there are distinctive mill designs, this report is not dealing with architectural designs of mills, except for the shape of the building around the equipment (step or flat). This report is primarily focused upon the engineering aspects of flotation milling. In this section, I will discuss the evaluation of significance and integrity of properties.

Guides

In order to understand guidelines suggested by preservationists, I referred to the following key sources: National Register Bulletin 15, How to Apply the National Register Criteria for Evaluation; National Register Bulletin 16A, Guidelines for Completing National Register of Historic Places Forms; and National Register Bulletin 42, Guidelines for Identifying, Evaluating, and Registering Historic Mining Properties.[1] Thus, the following section referring to the significance of historic properties has been summarized from the above bulletins. For more information on preparing National Register nominations, refer to these bulletins.

Establishing Significance of Milling Properties

The National Register of Historic Places requires an explanation of significance for a property under four criteria:

A. history (broad events or patterns)

B. people (historically important persons)

C. architecture (distinctive characteristics of a type)

D. archeology (information potential)

While this report has developed a context for the technological evolution of selective flotation, a milling property's significance does not have to be related to that particular context. A mill may be significant under any of the four criteria established by the National

Register of Historic Places, in any of a number of contexts.

Criterion A

Criterion A demands that a property be associated with "events that have made a significant contribution to the broad patterns of our history." Areas of significance could include business (such as the development of an important business associated with the extractive industry), commerce (the milling operation's production of a concentrate for commercial sale or sale to a smelter), community planning and development (the community that supplied the labor for the mine and mill), and engineering (the development of processes by mining engineers). For a listing of other areas of significance, refer to Bulletin 42, pages 15 and 16 and the master list in Bulletin 16A.

Criterion B

Criterion B states a milling property can possess significance if it is "related directly to a historically significant person." Examples could include James Hyde's career with Butte and Superior Mining Company developing a froth flotation unit at Basin and later the Timber Butte Mill, Arthur F. Wilfley's design and development of the Wilfley table for concentration, or Hezekiah Bradford's experimentation with the early skin flotation process in concentration.

Criterion C

Under Criterion C, a milling property may possess significance if it embodies the "distinctive characteristics of a type, period, or method of construction, or represents the work of a master, or possesses high artistic value, or represents a significant and distinguishable entity whose components may lack individual distinction." Criterion C is used for architectural characteristics of buildings, but it can also be used for significance of engineering structures. Innovations in the use of metal and concrete have been recognized in the realm of mining. Structures, such as the steel and concrete aerial tram platforms designed by Fred Carstarphen for the Shenandoah-Dives Mill, could be recognized in this category. In addition, mining engineering has played an important part in the progress of technology in crushing and concentration of ore. Arthur Wilfley's rippled table for concentration of minerals had significant impact on the development of milling in the twentieth century.

Criterion D

Under Criterion D, the historic archeological remains of a milling property may be significant if they have yielded, or may be likely to yield, important information. Eligible resources outlined by Bulletin 42 on page 17 include surviving equipment, landforms such as mill tailings, building foundations, roads, and train rails. Because archeological remains of milling operations are much more extensive throughout the United States than intact historic buildings and structures, this is an important criterion for evaluating milling properties. The information that can be gleaned from these sites are determined by "developing research questions, identifying the data requirements, and assessing the property's information content."

Establishing Integrity of Milling Properties

Integrity is the ability of a property to convey its historic significance to meet National Register of Historic Places requirements. It is important, however, to understand that integrity of a milling property cannot follow the same guidelines used for evaluating a historic building. Milling history was not static. Boom and bust cycles took a toll on properties related to the history of mining and milling. Properties were alternately developed, abandoned, and reworked.

In the boom-bust cycle of mining, mill properties were re-opened, updated, and modified to represent change in technology. Modification to a mill property might include the addition of new technology or alterations in a circuit to process either a new

mineral or several minerals. For instance, the Valmont Mill in Boulder County, Colorado was initially built in 1935 as a 100-ton gold and silver mill. In early1940, a market for fluorospar prompted General Chemical Company (later Allied Chemical and Allied Signal) to enlarge and reconfigure the mill to a one-circuit facility for the separation of fluorspar from custom ore. By the late 1970s, the mill had been reconfigured again to mill tailings for gold. A mill reflecting the evolution of technology can also retain a high degree of integrity. In this case, Valmont Mill's original flow pattern, even though modified temporarily, has retained its historic integrity and displayed the evolution of the mill building.

Seven Aspects of Integrity

The National Register recognizes seven aspects that define integrity: location (original), design (relatively unaltered), setting (reflecting its history), materials (original), workmanship (original), feeling (isolation, abandonment, or activity) and association (reflects historic association with mining and/or milling). In Bulletin 42, the authors noted that although individual components may lack distinction, the combined impact of separate components may illustrate the "collective image of a historically significant mining operation."

The important principle to consider when reviewing a milling property is the degree of integrity of the overall milling site. Acceptable levels of integrity may vary depending upon the significance considered. In Bulletin 42 (page 21), three examples for applying integrity standards to mining properties are summarized. For our purposes, the following examples can be used to describe the relative integrity of selective flotation mill sites that have engineering/technological significance. Example one involves a rare case where a mill site is intact. This mill, which would be determined to have an excellent level of integrity, consists of a complete milling system including conveyances to and from the site, arrival terminals and ore bins, crushing plant, mill building with equipment and flotation circuits intact, administrative offices, workshops, tailings site, and other aspects of the overall system. A second example notes a property that lacks visible buildings or has partially intact buildings.

However, key aspects of the site represent its association to flotation milling. This property would have a relative degree of integrity because the key components remain visible. In the third case, buildings are extant; but the inside of each building has been totally altered and key components have been destroyed in modern development. This property would have lost its integrity in regards to technological significance. Further discussion and examples can be found in Chapter 6.

Endnotes

1. U. S. Department of Interior, National Park Service, A Guideline for Identifying, Evaluating, and Registering Historic Mining Properties by Bruce J. Noble and Robert Spude. National Register Bulletin 42. (Washington, D. C.: Government Printing Office, 1992); U. S. Department of Interior, National Park Service, How to Apply the National Register Criteria for Evaluation. National Register Bulletin 15. (Washington, D. C.: Government Printing Office, 1990); and U. S. Department of Interior, National Park Service, How to Complete the National Register Registration Form. National Register Bulletin 16A. (Washington D. C.: Government Printing Office, 1991).

Chapter 6 - Results and Recommendations

The West was once dotted with hundreds of mining and milling operations. Most of what remain are abandoned sites, open shafts, and scarred land. The exact number of abandoned properties is difficult to gauge since there has never been a comprehensive inventory of properties. After their mines played out, mining and milling complexes typically became empty and desolate structures, ruins, holes, or tailing heaps. In the January 1899 <u>Mining Directory of San Miguel, Ouray, San Juan and La Plata Counties</u>, it was reported that in Baker's Park, San Juan County, Colorado there were 268 mill sites on record and 84 patented mill sites. Only ruins and historic archeological sites remain of these mill sites.[1] A 1980s study by mining engineer Harrison Cobb found a similar situation in Boulder County, Colorado. According to Cobb, Boulder County, although not a major mining district in Colorado, once had more than one hundred mills. Of those, there were only ruins, foundations, two deteriorating structures, and five operable mills in 1980.[2] The status of Boulder County's mines and mills are typical of those throughout Colorado, the West, and the United States.[3]

The primary purpose of this project was to identify extant resources of this vanishing property type and to evaluate them in terms of their integrity. Specifically, I sought to determine how many selective flotation mills (500-2,000 ton) are still in existence, and have a high level of integrity.

It should be noted that my research focused on the technical evolution of the flotation process and its equipment (engineering). As a result, the integrity of the sites was based upon the significance of the extant engineering equipment, not the mill buildings. In addition, I did not evaluate each of the mills for their historical significance, but only for their integrity related to the flotation process.

Also the scope of this project did not allow for me to visit and evaluate the historical significance of each of the mills listed. I relied on experts in the field of mining to assist me in making my determinations.

Rating Integrity

For the purposes of this study, I used the guidelines for integrity discussed in Bulletin 42, as mentioned in the previous chapter, to develop the ratings below. In terms of evaluating integrity, ratings for this report were assigned as follows:

- Excellent: The site displays a significant number of the buildings or processes listed in the property type section; retains a mining setting, possibly an abandoned or isolated location; reflects an industrial atmosphere appropriate to the period of significance; and has elements of the flotation process that are intact with little or minimal alterations.

- High: A majority of the property types are visible, displaying its association with milling. Also, the complex retains a mining setting; and reflects an industrial atmosphere appropriate to the period of significance. The majority of elements of the flotation process are intact, but may or may not be operational.

- Low: The mill building or buildings display integrity but the interior equipment has been drastically altered. Ancillary buildings are gone or drastically altered. Site may lack association with its original design.

- Poor: Remains are visibly in poor condition (falling down) or only foundations remain. Mill buildings have been moved

from their original location. The elements of the flotation process have been removed or destroyed. (Relocated properties generally do not qualify for National Register.)

- Historic Archeological Site: Remains are gone but the landscape reflects association to a mill site. The site may lack its above ground remains, but buried historical archeological deposits are highly likely to be present. The system (site dynamics, size, and location) surrounding mill technology is of crucial importance to the archeologist.

In the end, my survey research effort located only two exceptionally intact examples of selective flotation mills: the Ely Valley Pioche Mine and Mill, at Pioche, Nevada and the Shenandoah-Dives Mill, Silverton, Colorado, both of which reflect the distinctive characteristics (to be discussed) of flotation milling processes in early twentieth-century America.

The Shenandoah-Dives Mill site had all its original elements intact and was in excellent condition. The Ely Valley Pioche mill was found to reflect a high degree of historic integrity.

48

Identified Existing Properties

As discussed earlier in this study, flotation mills are designed and built according to natural features of the mill site and the configuration of the equipment used in the flotation process. Generally, the sizes of historic flotation mills fall into two categories: large (processing 500-2,000 tons of ore a day) and small (processing 50-200 tons per day).

HISTORIC MILL PROPERTIES						
SIZE OF MILL AND NAME	INTEGRITY AND CONDITION OF MILL SITE					
	Excellent	High	Low	Poor	Historic Archeological Site	Removed
Large Mills (500-2000 Ton)						
Shenandoah-Dives	X					
Ely Valley Pioche		X				
Idarado			X			
Independence			X			
Combined Metals				X		
Sunnyside					X	
Tintinc Standard					X	
Bunker Hill						X
Small Mills (50-200 Ton)						
Valmont		X				
Charter Oak		X				
Clayton		X				
Slowey		X				
Comet		X				
Timber Butte				X		
Unknown Size						
Juniata						X

LARGE MILLS (500-2,000 Tons Per Day)

This survey located five extant (500-2,000 ton per day) mills in a variety of conditions. Only one mill was in excellent condition, the Shenandoah-Dives Mill in Silverton, San Juan County, Colorado. One mill had high integrity, the Ely Valley Pioche Mill in Pioche, Lincoln County, Nevada. Three are in the low or poor categories. Other non-extant mills listed are examples of archeological and removed.

Excellent Integrity

Shenandoah-Dives Mill Silverton, San Juan County, Colorado

The Shenandoah-Dives Mill is included within the Silverton National Register Historic District (NR97000247) in Silverton, Colorado. Constructed in 1929, the Shenandoah-Dives Mill was designed for milling metals from low-grade gold ore using alkaline reagents in separation processes. The mill processed from 750-1,000 tons of ore per day in a multiple circuit plant that milled five metals: copper, lead, zinc, gold

Shenandoah-Dives (Mayflower) Mill, Silverton, Colorado, 1997. The addition of the crushing plant is visible behind (to the left) the office building (at the far right edge of the photo). From the crushing plant, the ore traveled up the conveyor to the top level of the mill where it was further crushed in the rod mill addition (visible on the left of the mill).

and silver. In operation until 1992, the Shenandoah-Dives Mill contains virtually all of its working components enclosed within a 1,000-ton mill complex. The mill complex includes a mill building that encompasses the main mill, a conveyor, crushing plant, tram terminal, steel rod mill, and workshop. The mill complex also includes a water storage tank, coal storage bin, office/assay building, guard shack, electrical transformer, lime storage building, and decantation and tailings ponds. An aerial tramway with numerous ore buckets hanging from the lines connects the mill to the Shenandoah-Dives Mine. An avalanche deflector at the uppermost end of the tramway is extant. The Shenandoah-Dives Mill complex is an excellent representation of 1920s-era milling technology in its original processing format. Since its construction, only minor changes (in an effort to increase production capacity) have been made to the mill facility (see Appendix A). The mill has been donated to the San Juan County Historical Society to be used as an interpretative center for flotation milling in the Rocky Mountains.

High Integrity
Ely Valley Pioche Mill Pioche, Lincoln County, Nevada

Ely Valley Pioche Mine and Mill of Pioche, Lincoln County, Nevada is a 500-1,000 ton-per-day, selective-flotation mill. When an early 1920s mill burned down, a second plant was built in the late 1920s and early 1930s on the partially inclined site above town. First powered by steam, the stack for the current mill is visible from the nearby town of Pioche. Electricity was run to the mill site at a later date. The ore (lead-zinc, gold, silver, and minimum copper amounts) from the Pioche Mine was hauled from the mine to the tram terminal by a small steam locomotive. Once loaded into the tram's ore buckets, the ore traversed approximately one and one-half miles to the mill at the upper end of the town. After the ore was processed in the mill, the concentrate was transported by the Union Pacific Railroad to Salt Lake City, Utah. The mill ran from the 1930s into the 1950s when it was eventually mothballed. It was briefly reopened in the 1980s. Currently, the mill is mothballed with its remaining equipment intact. One can view the ball mills, agitators, float cells, redwood tanks, and classifier. The tailings are piled nearby. Ore buckets are still visible, hanging from the tramline. However, changes in mill flow and loss of original equipment result in a rating of high integrity rather than excellent. The owner has expressed no desire to seek National Historic Landmark designation of this property.

Low Integrity

Idarado near Telluride, San Juan County, Colorado

The Idarado Mill (approximately 1,000 tons) was built by the Smuggler Union Mining Company in the 1920s near Telluride, Colorado. Subsequent operations have drastically altered the flotation circuits of the mill, but the mill building retains its historic integrity. Additional alterations at the mill site have prompted a rating of low historic integrity.

Independence Mine and Mill, Talkeetna Mountains, W. Palmer, Alaska

The Independence Mill site is included in the National Register property (NR74000440) and the Independence Mine State Historic Park. While buildings from the mill site are standing, the mill itself is gone. Constructed in 1937, the Independence Mill complex included several buildings: an office, bunkhouse, warehouse, mess hall and kitchen, commissary, work shop, welding shop, power and machine shop, and ore sorting plant. Remnants of the aerial tram are visible from the site to the mine. The association of the site to mining and milling activity is strong. Remaining buildings have been restored, but no longer representative of their historical significance. They are being used as administrative offices and a museum. Under the rating integrity assigned in this report, the site would be considered low due to the missing mill building and equipment.

Poor Integrity

Combined Metals Smelter and Mill Castleton, Nevada

Constructed about 1939-41, the Combined Metals Mill (1,500 ton) was a partially inclined and flat mill site. Active during the Second World War, the mill processed five mineral concentrates. Intermittent openings and closures led to its final closure in the 1980s. Equipment and materials were salvaged. The remains of the building are in general disrepair and the site is in poor condition.

Historic Archeological Sites

Archeology is concerned with past human behavior and contains information important in prehistory and history. We determine this through the remains of material culture, including artifacts, and features. Since most of archeological history tends to be buried, sites would require archeological investigations to determine their integrity.

Sunnyside Mill, Eureka, San Juan County, Colorado

The perfection of selective-flotation on complex ore bodies was attributed to the Sunnyside Mill, which is located near the National Historic Landmark-mining town of Telluride, Colorado. The mill at its peak was enlarged to process more than 1,100 tons of ore per day. After thirteen years of activity, the frame building succumbed to fire. The Sunnyside Mill is an example of a historic archeological site.

Tintic Standard Reduction Mill near Park City, Utah

A Historic American Engineering Record (HAER No. UT-12) team, including archeologists, historians, and architects, documented the Tintic Mining District. The most documented of flotation millsites; the Tintic HAER documentation includes four drawings, thirteen photos, and thirteen pages of documentation. The Tintic Standard Reduction Mill Site (NR78002700) has also been listed on the National Register of Historic Places. The flotation mill (1,000-1,500 ton), locally referred to as the Harold Mill, was built in 1920-21 to process copper, lead, zinc, gold and silver. Once an active mining district in the state and region, the Tintic Mining District was eventually abandoned around 1957. At the time of the HAER crew's visit, the mill site and leaching tanks were all that remained. Salvaging and weather conditions had reduced the site to rubble, foundation walls, and steps in the terrain.

Removed

Bunker Hill Lead Smelter Ore Concentrator and bins located in Shoshone County, Idaho

Prior to the Superfund cleanup of the Bunker Hill site, a HAER team documented the Bunker Hill district (HAER No. ID-29). A 1930s-era flotation mill, the Bunker Hill Concentrator processed approximately 1,000 tons per day of lead. At the time of the HAER documentation, the site was in disrepair, with equipment removed for salvage. The site has been reclaimed in a Superfund cleanup.

SMALL MILLS (25-200 Tons Per Day)

Although the goal of this survey was to locate large mills, information on smaller mills was discovered in the process. This survey found that small industrial size plants processing multiple metals were almost non-existent. I found a number of smaller single-circuit plants (50-200 ton) in the West. A more comprehensive inventory of these properties should be done. The following mills are only a sampling, but five of these sites reflect a high degree of historic integrity. The Clayton Mill (Clayton, Idaho), Valmont Mill (near Boulder, Colorado), Comet Mill (near Basin, Montana), and the Slowey Mill (near Superior, Mineral County, Montana) were 100-200 ton-per-day processing facilities. One 50-100 ton-per-day mill, the Charter Oak Mill (near Butte, Montana) was located. These extremely rare intact examples of twentieth-century mills represent the range in size and evolution of milling during the first half of the twentieth century. The following list is by no means comprehensive, and a comprehensive inventory should be compiled of the single-circuit mills.

High Integrity

Valmont Mill near Boulder, Boulder County, Colorado

The Valmont Mill near Boulder, Colorado was built in 1935 by the St. Joe Mining and Milling Company. The 100 ton-per-day mill was built to process gold ore from the Grand Republic Gold Mine at Chrisman, Colorado. In the 1940s, General Chemical Company (later Allied Chemical and Allied Signal) enlarged and reconfigured the mill to process fluorspar. Fluorspar was mined from the Invincible Mine near Jamestown, Colorado. From the mill, rail transported the product to Bayport, California where it was processed into hydrofluoric acid. The mill is a gravity-fed mill with water pipes drawing water up the hill to the site for use in its flotation process. Surfactants were added to the water to aid the minerals in adhering to the bubbles, where

Valmont Mill

they were swept off into the drying process in drum cylinders. At one time, the tailings were run underground to a tailing pond a quarter mile away.

Charter Oak Mill in the Elliston Mining District, near Helena, in Palo County, Montana

In the early 1930s, James Bonner leased the Charter Oak Mine and Mill property. After removing a small-frame mill, Bonner erected a gravitational, selective-flotation mill on the slope of the mountain. Over the course of time, a total of seven buildings were erected on the site. They included the mill building, compressor building (power), storage, lab, bunkhouse, and office. The wooden-frame mill, located near the Little Blackfoot River, processed approximately 50-100 tons of ore per day from the Charter Oak Mine, as well as ore from other mines in the Elliston Mining District. In 1943, the mill processed a total of 2,000 tons of ore for four principal metals: gold, silver, lead and zinc. There are remnants of different cell circuits located on the site, but the mill appears to have operated with one circuit of flotation cells. One-circuit operations were adjusted periodically to recycle tailings through the circuit. The process could be repeated five times with a different chemical bath each time to "float out" the desired mineral: gold, silver, lead, zinc and copper.

The operation has been used intermittently, but its peak years were during World War II. Presently abandoned, the mill is located on National Forest Service lands. Due to its remote location and being surrounded by gated fences, the Charter Oak Mill represents a high standard of integrity with a majority of milling and crushing equipment intact.

Clayton Mill Clayton, Custer County, Idaho

Constructed in early 1930s, the Clayton mill is a 100-200 ton per day processing facility. Its single circuit capacity milled hard-rock minerals from area mines.

Slowey Mill near Superior, Mineral County, Montana

Constructed in the 1930s, the frame and galvanized steel mill processed 100-200 tons per day through its selective-flotation process for the Nancy Lee Mine. Ore was trucked in from mines in the Keystone and Flat Creek area. Ore was crushed in a grizzly crusher traveling by conveyor where it was also hand picked before arriving at the mill plant. Most of the physical plant is intact, including partial remains of the crushing building, the mill building (intact), office, sorting belt, ore bins, water tank, tailings, outhouse, rail evidence and interconnecting roads.

Comet Mill at the Comet Townsite near Basin, Montana

Located in a historic mining town (now a ghost town) dating from the 1880s, the Comet Mill was erected in the 1930s. The frame and corrugated metal building houses a 200-ton per day operating mill. The Basin Montana Tunnel Company erected the mill, possibly on a former mill site, connecting it to an aerial tramway that traveled over the mountains to a smelter in Wickes. The mill closed in 1941 and its machinery was dismantled and removed. The ghost town has 88 structures related to historic mining ventures from 1869-1941.

Poor Integrity

Timber Butte Mill Ore Bins, near Butte, Silver Bow County, Montana

The Timber Butte Mill (approximately 100 ton) was the first American flotation plant. James Hyde moved his experimental plant from Basin to Timber Butte in 1912. Finished in 1914, the mill consisted of three separate buildings connected by conveyors and concentrate launders. A train to the mill site transported extracted ore. A small locomotive transported the ore cars from the main rail line to the mill site. The frame building with concrete foundation has been removed. The ore bins are the only remains of the historic mill today. The bins have been converted into a private residence. This site is rated poor.

UNKNOWN SIZE

Juniata Mill located in Mineral County, Nevada

The Juniata Mill is an historic archeological site with partial remains. The size of the processing plant has not been determined at this time.

A HAER team is currently documenting the Juniata Mill. The flotation mill was built about 1939 and operated until the 1950s when it was abandoned. The sides of the mill are still standing, but the roof has collapsed upon what remains of the equipment inside the mill. The site is in poor condition.

Endnotes

1. Mining Directory of San Miguel, Ouray, San Juan and La Plata Counties (Denver: County Directory Co., Jan.1899).

2. Harrison Cobb, Prospecting Our Past: Gold, Silver, and Tungsten Mills of Boulder County (Boulder, CO: The Book Lode, 1988), 5, 46-49, 137-141.

3. Ibid.

Chapter 7 - Summary and Conclusions

Flotation was the most important development of the twentieth century for the mining and milling industry. When the world had almost depleted its quantity of high-grade ore, flotation allowed for higher percentages and grades of concentrates to be collected from an abundant supply of low-grade ore. An integral part of the ore processing industry, flotation enabled millions of tons of rock to be processed to liberate the valuable metals within. However, the mills that represent the evolution of the flotation process are quickly being eradicated across the West. Representative and one-of-a-kind examples of twentieth-century mills have met their demise as a result of growth, technological advancements, weather conditions, salvaging and vandalism.

Preservation of selective flotation milling sites is at a critical juncture, with relatively few sites remaining of the hundreds once found in the United States in the early decades of the twentieth century. Only two intact 500-2,000 ton industrial metal mills remain in the American West. And there are approximately 25 small to medium mills that have survived, with most of those in poor and/or neglected condition. Given the significance of milling in the United States, additional inventory and evaluation of mining and milling sites needs immediate attention. While there are still extant properties in the West, the greatest challenge facing mining historians, preservationists, and cultural resource organizations is preserving them.

CASE STUDY: Technology of the Shenandoah-Dives Mill

Constructed in 1929, the Shenandoah-Dives Mill was designed for milling metal ores from low-grade gold ore (ore that is relatively poor in the metal for which it is mined) using alkaline reagents in separation processes. The primary metals processed at the mill were gold, silver, copper, lead, and zinc. In operation until 1992, the Shenandoah-Dives Mill contains virtually all of its working components enclosed within a 1,000-ton mill complex. The mill complex includes a mill building that encompasses the main mill, a conveyor, crushing plant, tram terminal, steel rod mill, and workshop. The mill complex also includes a water storage tank, coal storage bin, office/assay building, guard shack, electrical transformer, lime storage building, and decantation and tailings ponds. An aerial tramway connects the mill to the Shenandoah-Dives Mine. The mine will be capped and is not included, due to its lack of historical integrity, within the boundaries of the National Register district. An avalanche deflector at the uppermost end of the tramway is extant. The Shenandoah-Dives Mill complex is an excellent representation of 1920s-era milling technology in its original processing format. Since its construction, only minor changes have been made to the mill facility in an effort to increase production capacity.

The Shenandoah-Dives Mine and Mill are in the Animas Mining District of the San Juan Triangle. The mine is on the south slope of Little Giant Mountain in the section of the range referred to as King Solomon Mountain; the mine's portal is 11,200 feet above sea level (asl). The mill (9,700 asl) is two miles northwest of the mine, at the base of Arrastra Gulch near the Animas River. A 9,526-foot aerial tramway connects the mill complex with the Mayflower portal of the mine.[1] The Shenandoah-Dives Mill, which is locally referred to as the "Mayflower Mill" after the Mayflower portal of the mine, is two miles northeast of Silverton off of Highway 110.

The ore from the Shenandoah-Dives claim was originally crushed at the mine in an underground crushing plant that housed a Telesmith 16-A gyro primary crusher and a 4-foot Symons standard cone crusher. The ore was crushed to approximately _-inch-size gravel. Machines used for crushing are generally one of three types: reduction gyratory crushers, jaw or rolls. The gyro primary crusher can handle a large capacity of ore to be crushed. With the marketing of the Symons standard cone crusher, it quickly became the standard in the industry.[2] The crushed ore was then loaded into ore buckets at the upper terminus of the aerial tramway, and sent down the mountain to the Shenandoah-Dives Mill where it was unloaded into ore bins at the tram building. In 1961, a new crushing plant was constructed at the mill, and the short-head cone crusher was moved from the mine to the new plant. Up to this point, the mill had processed ore from the Shenandoah-Dives Mine, as well as custom ores from other mining companies in Silverton. Following the construction of the new crushing plant, the mill also processed ore from other mining operations in the area, including the Silver Lake lease and independently-owned mines in the area.

Aerial Tramway

The aerial tramway between the Shenandoah-Dives Mine and Mill was designed by Fred C. Carstarphen. At the time of its construction in 1929, it was the only tramway of its design in Colorado.[3] Most tram towers of that period were pyramidal, and made of timber or timber-and-steel components. By contrast, the Shenandoah-Dives tramway had rectangular, riveted-steel structures that could better withstand the snowfall and avalanches in the San Juan Mountains. Carstarphen's design also called for fewer and larger towers. When it was built, the Shenandoah-Dives aerial tramway was the longest tram line in operation in the San Juans, and could travel at a speed of 500 feet per minute.[4] The tramway carried men as well as ore. With only a steep, mountainous trail to the mine, miners used the bucket to travel to and from their jobs. (In emergencies, passengers in the ore buckets could pull a rope to activate a signal switch in the lower terminus, which, in turn, initiated an emergency response.)

As originally constructed, the 9,526-foot aerial tramway had 11 steel towers (Towers 1-11) and three steel, double-cable anchor stations (Stations A, B, and C). The anchor stations stabilized the tension of the cable that held the ore buckets. All of the anchor station towers are of rectangular shape with steel crossbars and corner lengths, with a ladder attached for maintenance. The towers have concrete foundations, and are of varying height, reflecting the contour of the mountain and the approximate 1,400-foot difference in elevation between the upper and lower terminal points. At one time, the towers also supported telephonelines between the mine and the mill.

In addition to the steel towers, the Shenandoah-Dives Company constructed wooden towers at the upper terminus (Loading Terminal) and lower terminus (Discharge Terminal). These two towers created a level plane for the arrival and discharge of ore buckets, as well as providing tension for the cables. The loading and discharge towers were referred to as terminals in day-to-day management of the aerial tramway. The Loading Terminal is no longer extant. The Discharge Terminal, located at the tramway's lower terminus, is a wooden, rectangular tower on a concrete base, and is similar in design to the steel towers. The aerial tramway system also included a lift tower near Anchor Station A. The lift tower raised the Shenandoah-Dives aerial tramway over the Iowa-Tiger aerial tramway, which ran counter to it. The lift tower no longer stands; Shenandoah-Dives Company tore it down c. 1942, following the abandonment of the Iowa-Tiger tramway.[5] In all, the Shenandoah-Dives aerial tramway contained 17 towers. Of these, 12 are extant: 8 steel towers, 3 double-cable anchor station towers, and one wooden tower (the Loading Terminal Tower).

The tramway's ore buckets traveled across a $1^3/_8$-inch stationary cable and a 7/8-inch traction cable. The traction cable was basically a wire rope with a fiber core with a greater degree of flexibility than cables.[6] Three double-cable anchorages are located at Anchor Stations A, B, and C. Cable anchorages are large concrete blocks with steel I-beams imbedded in them. Even after an avalanche took out an upper section of the aerial tramway in the mid-1960s, the cable anchorages held up the tension cable of the aerial tram and its buckets. By anchoring the segments, the tension cable was secured to accommodate the curvature of the slope and offset the possibility of losing the entire cable system in an avalanche.

The aerial tramway normally carried 52 numerically-assigned tram ore buckets, two timber carriers, and one automatic track oiler. Two sizes of buckets, 17 cu. ft., and 21 cu. ft., were spaced at intervals of 400 feet. The 21-cu.-ft. bucket had a round bottom and carried one ton of ore. The 17-cu.-ft. bucket had an angled bottom and carried 1,600 pounds of ore. The average payload was _ ton. All of the carriers were equipped with four wheels (referred to as 4-wheel trucks), instead of the usual two, which guided the ore buckets over the cables. When the tram was running at full speed, an ore bucket took 45 minutes to make a complete cycle from the mine to mill and back again.[7] Numerous ore buckets remain on the line.

The buckets on the cables were driven by two 50-horse-power General Electric motors that were originally located at the upper terminus of the tramway. One of the motors was a variable-speed type, which could reverse the tramway operation. The control of the bucket was maintained by two grip sheaves connected to the motors. At the mine's loading terminal, the bucket was chain secured under a loading chute. The operator pushed the loaded bucket out to the "gripper" to engage the bucket on the traction rope for its travel down the mountain. As the descending ore bucket activated the system, power flowed back into the line, and the motors controlled the downward speed, as well as supplying the power to draw the load up the mountain. As each ore bucket neared the mill's discharge terminal, the grip sheaves controlled the bucket's approach into the terminal building. The bucket was met by a waiting tram operator overseeing the bucket's arrival and diversion from the cable to a monorail for loading and unloading into the ore pocket. Each bucket attached or detached automatically from the traction cables at the terminals, and then was manually diverted by the operator to dump its load. The operator pushed the empty bucket to the exit side where the grip action was reversed and the car "gripped out" as it clamped onto the moving traction rope.[8] As the arriving bucket was diverted, a bucket on the exiting side was activated and began its slow ascent up the mountain.[9] At the upper terminus of the aerial tramway, a 300-foot auxiliary line (referred to as a stub tram) delivered supplies to the mine and boarding house from the main line. The auxiliary line no longer remains, nor do the main buildings at the mine area.[10] However, although the integrity of Shenandoah-Dives Aerial Tramway has been impacted by the downing of several towers, including the auxiliary line, the Aerial Tramway is an exceptional example of only a few remaining aerial tramways that have survived the harshness of high-altitude mountain climes.

Avalanche Deflector

The deflector is a massive, three-sided structure with walls of native rock and mortar; the cavity is filled with stone, dirt and sand. In 1938, an avalanche destroyed Towers 1-5, which the company rebuilt. In an effort to prevent future accidents of that type, the company also constructed an avalanche deflector above Towers 1 and 2. The avalanche deflector was designed by Charles Chase. According to Joe Todeschi, who helped construct the deflector, Chase designed a structure he believed would withstand the snowfall depths at 14,000 feet asl. In the region, "snow breakers," as they were referred to, were typically constructed of wood. Eventually, the wood would rot and then the weight of the snow would finish off the structure. As a result, Chase decided to build a stone structure, and hired Carlo Poloine, an Italian immigrant and skilled stone mason employed in the Shenandoah-Dives Mine, to build the "snow breaker."[11]

Poloine and four assistants constructed the deflector in the summer of 1938. Rocks gathered from the mountain side were used to build the deflector. The walls of the triangular structure consist of native rock mortared together. The lower wall of the triangular-shaped structure is the largest. The side walls were built to accommodate the steep terrain of the mountain; thus, those walls are shorter in height. Poloine began by smoothing the sides of the rocks the other men gathered. The rocks were smoothed on both sides so that the inner wall of the cavity of the triangular shape is as smooth as the outer. After the initial shape of the breaker was laid out, Poloine mortared each rock together to fashion the walls. The men used a single or double jack, long-handled hammers of various weights, to break up larger rocks into the size Poloine needed. As the lower wall was built, fill of rock, dirt and sand was thrown behind it; and Poloine worked smoothing rock for the wall on the growing pile of fill. The rock used in the

inner and outer walls were smoothed, even though the inner wall is not seen because of the fill of stone, dirt, and sand. According to Todeschi, "The snow breaker was built to last a lifetime."[12] The structure has a high degree of integrity, and appears to be unchanged.[13]

Shenandoah-Dives Mill

The Shenandoah-Dives Mill, a four-level frame and metal building with an irregular floor plan, is a truss-timbered framed building with both gabled and shed corrugated-metal roofs on its various levels. It is located approximately 100 feet north of Colorado Highway 110, the main route between Silverton and Eureka. The massive, sprawling mill building encompassed several milling and processing operations, and the building basically can be divided into the main mill, tram terminal, conveyor, crushing plant, steel rod mill, and workshop. All of these operations were interconnected and under one roof.

In 1929, the Denver-based company of Stearns-Roger Engineering designed and oversaw the erection of the gravity-flow mill and its equipment. The mill was built in 18 weeks from a prefabricated kit.[14] Arthur J. Weinig designed the metallurgical portion of the mill. The four-level building is terraced into the mountainside, which enhances the gravitational flow within the mill. The truss-timber frame of the building is constructed of Oregon fir. Native timber is used for sheeting. The gable and shed roof covering the four levels is corrugated, galvanized metal. The exterior was initially painted an aluminum color in 1932, but now has sections painted in aluminum, green, and rust-red. The foundations are concrete.[15] Over time, the mill was enlarged to a capacity of 1,000 tons. Its current dimensions are approximately 90 feet across the back and 106 feet across the front; the sides are approximately 252 feet. Metal and frame-sided additions reflect functional modifications made at various times, but the overall lines and shape of the plant have remained virtually unchanged since its construction in 1929.[16]

The mill is partially covered with corrugated-metal siding. The top or upper level has a gabled, corrugated metal roof. The second level has a corrugated-metal shed roof. The third level has a gabled corrugated-metal roof. The lower level has an aluminum-colored, corrugated-metal, shed roof. Initially, the mill had banks of windows at each of the four levels.[17] In 1981, metal siding was placed over the windows, at which time electrical lighting systems were enhanced to supply the needed light. On the north side, there are three 9/9 side-

by-side stationary windows, located on the third level. The east side contains two sets of triple side-by-side windows (9/9/9), as well as one 9/9 side-by-side set of stationary windows. On the south side of the building, 15 single-hung, 9-pane windows remain on the third level. On the west side, there are six 9/9 double-hung windows on the third level. Over 80 windows were in the mill when it was built in 1929. Six stationary 9-pane windows are in the conveyor house that runs from the crushing plant to the mill proper.

The tram terminal is a rectangular single-story, frame and metal building with a gabled corrugated-metal roof. It was originally built as a single-story, frame building in the side of the mountain. When the crushing plant was added to the mill, the rear of the tram terminal, left intact, was enclosed inside the plant. A truss-timber frame supports the front (northeast side), which includes the lower terminus opening for arriving ore buckets. Fred Carstarphen, who designed the tramway, designed the tram terminuses. Carstarphen also directed the 12-man crew that constructed the tram terminal. The rectangular building is approximately 28 feet x 60 feet and houses the tension equipment for the aerial tramway, the shop for tram repairs, and the loading end of the 200-foot ore conveyor to the mill. The building has corrugated metal siding. The north side retains its original set of four 6/6 double-hung windows. The roof is gabled and covered with corrugated metal.[18]

When the aerial tramway was functioning, stationary and traction cables ran approximately 9,526 feet from the lower terminus at the Shenandoah-Dives Mill to the mine's upper terminus building. A series of snow avalanches in the 1960s ripped down approximately 1,500 feet of the upper terminus' cable and Towers 3-5. At the time of the avalanches, the mill was no longer processing ore from the Mayflower Mine, and the line was not rebuilt.[19]

An attached frame-and-corrugated metal workshop is on the north side of the mill building. The workshop has overall dimensions of 24 feet x 44 feet with a corrugated metal shed roof and exposed rafters. The workshop was built in 1929 at the same time as the mill. It has eight 9/9 double-hung windows on the north side. The machine and welding workshop was used for maintenance on machinery within the plant, as well as fabricating engines and small locomotives for use in the mine. Flat cars on railroad lines hauled large pieces of equipment into the workshop for maintenance. The rail lines within the shop were 30 inches apart, matching the lines at the mine. The fabricated machinery

was transported to the mine via the aerial tramway.

Inside the mill plant, the original milling machinery remains virtually intact, despite periodic upgrades which include the addition of a rod mill and crushing plant within the mill complex. Most of the original machinery was used throughout the entire productive life of the plant and remains functional. The 1942 flow chart shown on page reflects the processes that were used in the mill throughout its operational history. In 1942, the mill was granted permission to continue production for base-metals by the War Production Board, who had closed plants processing gold or gold by-products. Typically, the plant would have processed a by-product gold concentrate, so the years before and after this flow chart would have included the gold concentrate circuit. The milling processes were as follows:

From the ore pocket, a 24-inch-wide pan conveyor moved the ore to a 250-foot gradient belt conveyor which transported the ore to a round, steel ore bin located within the mill. The 250-foot belt conveyor is housed in a gabled frame-and-corrugated-metal structure. From the 1200-ton ore bin (1), the rock traveled via a 24-inch-wide pan conveyor (2) to a four-foot Symons short head cone crusher (3) which dropped into the No. 86 Marcy

1942 Flow Sheet:

1. 1,200-ton ore bin
2. Pan conveyor, 24"
3. Symons short-head cone crusher
4. No. 86 Marcy grate ball mill
5. Dorr quadruplex classifier 12'x26'
6. No. 64 Stearns-Roger ball mill
7. Bucket elevator 35'x22'
8. Trash trommel, 9-mesh, 2-1/2'x6'
9. Belt elevator 24"
10. Three No. 6 Wilfley tables
11. 20-cell No. 21 M.S. flotation
12. 20-cell No. 21 M.S. flotation
13. Liberty Bell type sampler
14. Hydroseal pump, "B" frame size
15. Wilfley pump, 2"
16. Wilfley pump, 2"
17. Wilfley pump, 2"
18. Wilfley pump, 2"
19. Dorr thickener, 35'x10'
20. Stearns-Roger ball mill, 4'x10'
21. Esperanza-type classifier, 6'x16'
22. Wilfley pump, 3"
23. Wilfley pump, 3"
24. 8-cell No. 18 Denver flotation
25. Denver conditioner, 3-1/2'x5'
26. Wilfley pump, 2"
27. 6-cell No. 18 Denver flotation
28. Liberty Bell type sampler
29. Denver 1" concentrate pump
30. Dorr thickener, 35'x10'
31. Settling box, 3'x6'x2'
32. Dorr filter 2-1/2'x6'
33. Dorr filter 5'x10'
34. Table concentrate bin
35. Zn concentrate bin
36. Pb-Cu concentrate bin

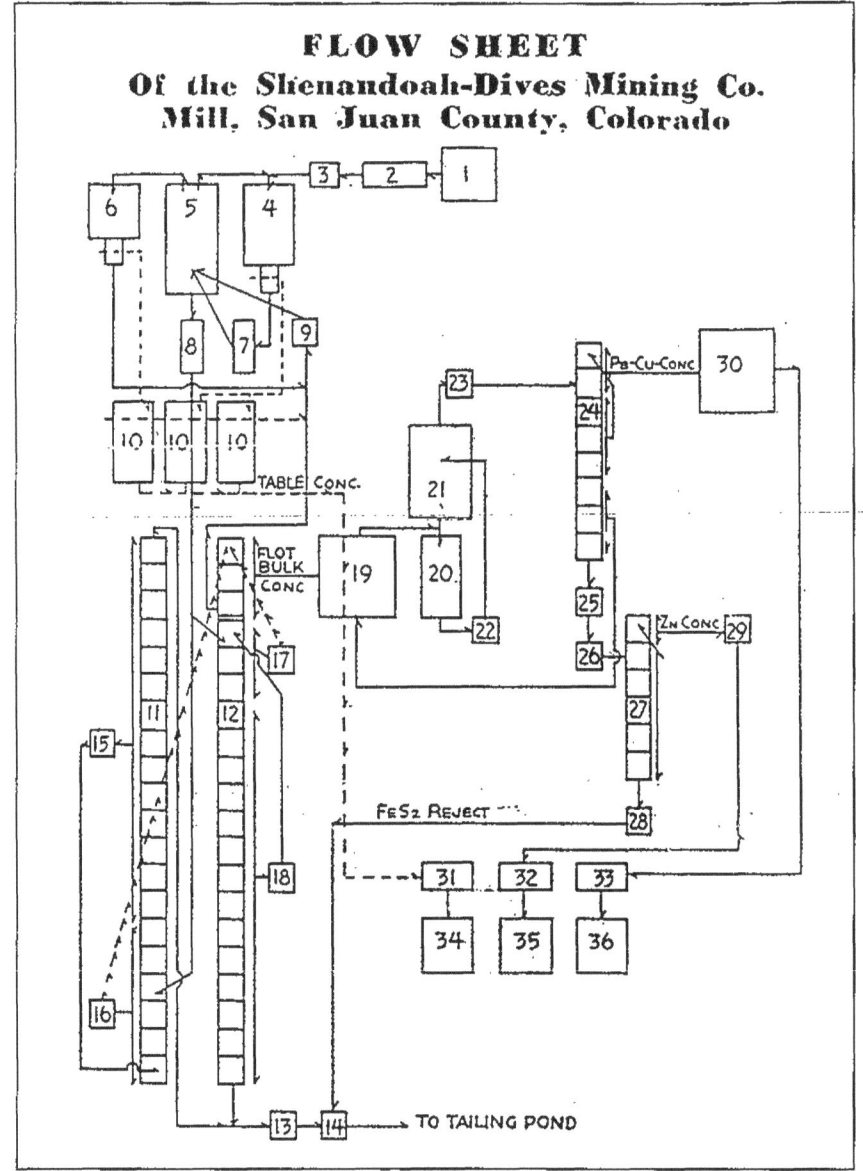

FLOW SHEET
Of the Shenandoah-Dives Mining Co.
Mill, San Juan County, Colorado

grate ball mill (4) where the ore was ground further with steel balls in the revolving center cavity. The Marcy ball mill was a wet-crushing mill with water being introduced to the crushed ore at this juncture.[20] At this stage, the product was a thin ore mud.[21]

The mud then traveled through a distributor (5), which is another "screening" process; trommel (8); another screen; and to three Wilfley tables (10). The Wilfey tables, activated by electricity, vibrated the muddy water introduced at the upper side of the table (feedside), separating coarser material as it flowed across the grooved table. The heavier, higher-grade concentrate (lead and gold) settled along the grooves at the bottom of the table (concentrate side). From the concentrate side of the Wilfley table, the heavy concentrates rode off the table edge into two settling or dewatering devices (19), such as a Dorr thickener. The thickeners (19) and settling tanks (31) removed the water from the mud. The thickener was a conical, settling tank with slow moving rakes. The coaser material settled into the bottom of the cone. The rake moved through the material causing lighter minerals to rise and wash off into a launder (filter) that carried the concentrate to a settling tank where a final drying process took place. The water was then filtered and returned for mill use. From the settling tank, the dry concentrate was scraped off and moved into a bin (34), to await transport.

The oversize of the gravel material then traveled by elevator to a 12-foot x 26-foot Dorr Classifier (5), where it was again washed and separated. The oversize was returned to the ball mill for further grinding, reprocessing, and screening (4, 6, 5, 8). The overflow continued on its way to the 40-cell bulk flotation circuit (11 and 12)).

In the flotation unit, the flow of mud went through three types of cells: cleaners, roughers, and scavengers. The cleaner cells had a mild concentration of reagents and short flotation time. The scavenger cells had a higher concentration of reagents and a longer flotation time. The water bath in the minerals-separation circuit was treated with an alkaline reagent. Individual reagents attracted particular minerals to the foam (froth) atop the water in the cells. Depending on the metals that were to be the end result of the process, different reagents were introduced to the process.

Entering through the feed inlet, each cell had an agitator and low-pressure blower that introduced a flow of air bubbles in the agitating compartment. The agitator blades stirred up the mix causing the froth to overflow and coarse sand to drop to the bottom. The coarse sand either returned through the grinding process or passed out as a tailing discharge. The concentrate, buoyant on the froth, was drawn off the top of the water in the froth-separating compartment.

The copper-lead concentrate from the first minerals separation flotation machine traveled to a Dorr thickener tank 35 feet in diameter (19). In the Dorr thickener, the concentrate was "de-watered" (the water returned to the mill) and sent on for crushing (20), classifying (21), and flotation (24). In the Dorr thickener, a thickener was added to the tailing to cause overflow of water for recycling. The concentrate traveled into a 5-foot x 10-foot Dorr filter (33). The filter was a drum that revolved, forcing the water out and the mineral to attach to its sides. The minerals were then scraped from the sides of the drum and moved to an 18-foot x 18-foot x 12-foot concrete concentration bin (36) to await transport to a smelter. (The Dorr filter was later replaced with a disc filter or "canvas wheel" that took up less space and processed a dryer concentrate.)

After passing through the Denver-built, minerals separation machine (24), the tailings were then zinc reactivated to pass through the Denver conditioner and Liberty Bell type sampler and flotation cells (25, 26, 27, 28) where the

product was separated into a zinc concentrate or a reject product of waste. After the zinc concentrate was separated, it was pumped (29) to a 2-1/2 foot x 10-foot Dorr filter (32) to be separated from the water. The dry concentrate was moved into concentration bin (35) to await transport to a smelter. The 1942 process changed over time with circuits added for processing a total of five concentrates, but it used basically the same technology and machines as listed on this flow sheet.

Basically, the flotation process cycled and recycled a muddy flow through grinders, sorters, separators, and dryers before ending up in concentrate bins awaiting shipments.

An inventory of the buildings in 1929 listed the mill, tram terminal with aerial tramway, sampling plant, office and assay building, electrical transformer, and two stave water storage tanks.[22] In 1937, a 6-foot x 5-foot ball mill was added, making a total of two ball mills. The plant processing capacity was then 700 tons.[23] In the 1960s, as the mill began to purchase more ore, the owners enlarged the main mill and added a zinc circuit for processing three concentrates, lead, zinc, and copper.[24]

In 1961, Standard Metals Corporation, which then owned

the mill complex, removed the mill's original sampling plant and replaced it with a new crushing plant. The crushing plant is attached to the mill building, and ties into the 250-foot conveyor system. The plant has a rectangular floor plan, approximately 40 feet x 60 feet, and a gabled corrugated-metal roof. Also at that time, the Standard cone crusher that had been at the mine was moved to the crushing plant. The short-head cone crusher that had been in the ball mill portion of the mill was also moved to the new crushing plant.

Despite these changes, the flow chart of the ore was only altered in the fact that the ore was now removed from the pocket ore bins, crushed in the plant, and then transported up the 250-foot conveyor to the mill. The rest of the process remained the same.

As technology advanced, Standard Metals Corporation continued to upgrade the operation. In 1975, a steel rod mill, which could grind a finer product, was added to the southwest corner of the mill. The steel rod mill, which was approximately 92-1/2 feet x 50 feet with a corrugated-metal shed roof, was added to the uppermost level of the mill in the southwest corner (where the two ball mills processed the ore). With the addition of the steel rod mill, the ore that had been previously crushed to

approximately ¾ inch was processed further, resulting in a fine grind about 1/8 to 1/10 inch in size.[25] It was also during this time that one of the mill's stave water tanks was torn down.

At the end of the milling process, the final tailings were pumped to tailings piles or slurried into the Animas River. In 1934, after researching methods that were believed to be environmentally sound, mill superintendent Charles A. Chase decided to use an innovative tailings pond method perfected and utilized by J.T. Shimmin in Butte, Montana. Altering this method to fit the Shenandoah-Dives Mill's specific needs and terrain, the Shenandoah-Dives Company began depositing its surplus into tailings ponds south of the mill. A chute or tailings flume delivered the pumped tailings to the pond area. At the time, the utilization of tailings ponds was atypical for the mining industry as a whole. Generally, environmental concerns were not at the forefront of the industry's interests; profitable veins of ore were of greater concern. As a result, the Shenandoah-Dives Mill was one of a limited number of mining enterprises that employed environmental, as well as cost-efficient, methods in their day-to-day activities.[26]

A V-shaped box delivered the tailings to the pond area. The box was made of two 2-inch planks;

one was 12 inches wide, the other was 10 inches wide. The box was supported by a 20-foot-high trestle that was set on gradient to initiate flow of the tailings. Upon arrival at the pond, the tailings were distributed by a grooved 20-foot-long board to form a "wall of sand" in the shape of pond. The technician would move the board periodically to retain a level top to the pond. In order to draw off water without stirring up the sediment, a wooden box was laid in a trench up the hillside prior to depositing the tailings. The top of the box had a series of holes 1-1/2 inch in diameter. As the water level rose, the lower hole was corked off to elevate the water level. As each subsequent hole was reached, a cork was placed in the hole. Once an established level of water was attained, the water was drained off through pipes into a decantation pond located on a lower plane than the tailings ponds. In practice, only a small amount of water was actually decanted or lost through evaporation. The hillside absorbed the greater volume.[27] As one pond filled to capacity, the flume was lengthened and another pond was begun. At the Shenandoah-Dives Mill, the ponds filled a triangular shape, following the mill's property lines. Four ponds are located within the triangle. The two oldest ponds, #1 and #2, are from the 1930s-50s; while #3 and #4

ponds were created in the 1970s-90s.

Endnotes

1. "Report to the Bureau of Mines, State of Colorado for the year 1930: Metals, Mines, and Mills," submitted by Shenandoah-Dives Syndicate for the Mayflower Mine, (County: San Juan; Dist.: Animas), 15 August 1930, (Photocopy).

2. Robert Richards, Textbook of Ore Dressing (New York and London: McGraw-Hill Book Co., Inc., 1940), 19-20 and A. M. Gaudin, Principles of Mineral Dressing (New York and London: McGraw-Hill Book Co., Inc., 1939), 35-36.

3. "Report to the Bureau of Mines, State of Colorado for the years of 1929 and 1949: Metals, Mines, and Mills," submitted by Shenandoah-Dives Syndicate for the Mayflower Mine, (County: San Juan; Dist.: Animas), 6 December 1929 and 8 February 1950; T. R. Hunt and Leon M. Banks, "Operation of the Shenandoah-Dives Mining Company," The Mines Magazine, February 1934, p. 49 ; William Jones, interview by Dawn Bunyak, Silverton, Colorado, 15 June 1995; and Photographs of the Aerial Tramway taken in the early 1930s. Western Historical Collection of the Denver Public Library, Denver, CO (notations on the back of the photographs by the photographer).

4. "Report to the Bureau of Mines, State of Colorado for the year 1929 and 1949: Metals, Mines, and Mills;" Fred Carstarphen, "The Mayflower Aerial Tramway of the Shenandoah-Dives Mining Company," The Colorado School of Mines Magazine, XX no. 9 (Sept. 1930): 24-26; Margaret Barge, "History of Shenandoah-Dives Mining Company", 1959(?) TMs [photocopy], p 6, San Juan

County Historical Society, Silverton, CO; to [Dawn Bunyak, National Park Service, Denver, CO],TLS, 8 August 1995.

5. William Jones, correspondence to Dawn Bunyak, 8 August 1995.

6. Darnall Zanoni, correspondence to Dawn Bunyak, 1 April 1996.

7. Ibid.

8. William Jones, correspondence to Dawn Bunyak, 1 April 1996 and Darnall Zanoni, correspondence to Dawn Bunyak, 1 April 1996.

9. "Report to the Bureau of Mines, State of Colorado for the year 1929 and 1949: Metals, Mines, and Mills"; Barge, "History of Shenandoah-Dives," 6; Hunt, "Operations of Shenandoah-Dives," 49; "Shenandoah-Dives," Mining World, 3-7; Jones, Interview, 15 June 1995; Photographs, Western Historical Collection; Chase Collection of Correspondence [photostat] in William Jones' Private Collection, "Charles A. Chase wrote to Fred Carstarphen on 14 September 1930 concerning the proper name for the tramway for publication purposes was to be referred to as the Shenandoah-Dives Aerial Tramway." William Jones, Correspondence to Dawn Bunyak, 3 August 1995.

10. "Shenandoah-Dives," Mining World, 3-7; William Jones, correspondence to Dawn Bunyak, 1 April 1996; Darnall Zanoni, correspondence to Dawn Bunyak, 1 April 1996; and Jones, Interview.

11. Carlo Poloine was the Italian spelling of his surname; but it would eventually become the Americanized "Palone," according to Joe Todeschi in a phone interview with Dawn Bunyak on December 16, 1996.

12. Joe Todeschi, phone interview with Dawn Bunyak, 16 December 1996.

13. William Jones, phone interview with Dawn Bunyak, 27 July 1995; Frederic J.

and William Jones, Montrose, Colorado, Athearn, Historian, phone interview with Dawn Bunyak on 6 September 1995; Frederic Athearn, photograph of Mayflower Mine--Avalanche Diverter and Tram Tower No. 1 taken in 1991. "It is important to note the upper terminal is still intact (along with towers 1 and 2) and is complete with all mechanical equipment, chutes, grip sheaves, gears etc. This site is located on the Mayflower #2 lode mining claim U.S. Mineral Survey Number 16551," William Jones notes in his correspondence of 1 April 1996 and 21 April 1996 to Dawn Bunyak.

14. William Jones, phone interview by Dawn Bunyak, 18 August 1995 and Jones, interview, 15 June 1995. Due to the evidence of archival photographs and speed in erection of the mill, it is surmised that the mill was designed and pre-cut in Denver, bundled up, and shipped to Silverton for erection by the construction crew.

15. Jones, interview; Allen Nossaman, personal interview by Dawn Bunyak, 13-15 June 1995; "Report to the Bureau of Mines, State of Colorado for the years of 1927, 1929, 1930: Metals, Mines, and Mills," submitted by Shenandoah-Dives Syndicate for the Mayflower Mine, (County: San Juan; Dist.: Animas), 8 February 1928, 28 May 1929, 6 December 1929, and 15 August 1930.

16. Nossaman, interview; Report to the Bureau of Mines, State of Colorado for the years 1929-1930: Metals, Mines, and Mills, submitted by Shenandoah-Dives Syndicate for the Mayflower Mine, (County: San Juan; Dist.: Animas), 28 May 1929, 6 December 1929, and 15 August 1930; and Colorado Historical Society Historic Building Inventory Record filed by Allen Nossaman on the Mayflower Mill, 1995.

17. Mayflower Mill of the Shenandoah-Dives Mining Company in Silverton, San

Juan County, Colorado, Photographs, Western History Department, Denver Public Library, Denver, Colorado.

18. "Report to the Bureau of Mines, State of Colorado for the years of 1929 and 1949: Metals, Mines, and Mills," submitted by Shenandoah-Dives Syndicate for the Mayflower Mine, (County: San Juan; Dist.: Animas), 6 December 1929 and 8 February 1950; Barge, "History of Shenandoah," 6; Hunt, "Operations of Shenandoah-Dives," 49 and "Shenandoah-Dives," Mining World, 6.

19. "Report to the Bureau of Mines, State of Colorado for the years of 1929 and 1949: Metals, Mines, and Mills"; Barge, "History of Shenandoah," 6; Hunt, "Operations of Shenandoah-Dives;" "Shenandoah-Dives," Mining World, 7; Jones, interview; and Jones, correspondence, 3 August 1995. There has been a discrepancy in the reporting of the length of the aerial tramway cable in a variety of publications. The figure referred to in this report is drawn from the survey notes compiled by Fred Carstarphen for designing and erection of the aerial tramway. These survey notes were supplied by William Jones of Montrose, Colorado who owns photostat of correspondence collection of Charles A. Chase, manager and Vice-President of the Shenandoah-Dives Mining Company.

20. A. M. Gaudin, Principles of Mineral Dressing (New York and London: McGraw-Hill Book Co., 1939), 94-97 and T. R. Hunt and Leon M. Banks, "Operations of the Shenandoah-Dives Mining Company," The Mines Magazine (February 1934): 51.

21. Paul W. Thrush, A Dictionary of Mining, Mineral, and Related Terms (Washington: U. S. Dept. of Interior, 1968), 1026.

1930 and 1932, Mining Collection,

22. "Report to the Bureau of Mines, State of Colorado for the year 1929: Metals, Mines, Mills," submitted by Shenandoah-Dives Syndicate for the Mayflower Mine, (County: San Juan; Dist.: Animas), 28 May 1929.

23. "Report to the Bureau of Mines, State of Colorado for the year 1937: Metals, Mines, Mills," submitted by Shenandoah-Dives Syndicate for the Mayflower Mine, (County: San Juan; Dist.: Animas), 5 January 1938.

24. William Jones, correspondence to Dawn Bunyak, 1 and 21 April 1996.

25. "Report to the Bureau of Mines, State of Colorado for the year 1962 and 1975: Metals, Mines, Mills," submitted by Shenandoah-Dives Syndicate for the Mayflower Mine, (County: San Juan; Dist.: Animas), 21 January 1963 and 15 December 1975; Jones, interview; Darnall Zanoni, personal interview by Dawn Bunyak, 13-15 June 1995; Nossaman, interview; and photographs, Western History Collection of the Denver Public Library, Denver, Colorado.

26. Duane A. Smith, Mining America: The Industry and the Environment, 1800-1980 (Kansas City: University of Kansas Press, 1987), 34-41.

27. Charles A. Chase and Dan M. Kentro, "Tailings Disposal Practice of Shenandoah-Dives Mining Company," in Mining Year Book, Mining Congress Journal, (Denver, CO: Colorado Mining Association, 1937), 31-32 and 74-76; and Charles A. Chase and Dan M. Kentro, "Tailings Disposal Practice of Shenandoah-Dives Mining Company," Mining Congress Journal 24 (March 1938): 19-22. In this article, they argue the cost efficiency, as well as environmental benefits, of using tailing ponds instead of other methods of disposal.

CASE STUDY:

History of the Shenandoah-Dives Mining Company

Between 1882-1918, the San Juan Triangle district in the Silverton area of southwest Colorado produced more than $65 million in the precious metals gold and silver. After World War I, mining in the Rocky Mountain West suffered a severe decline. However, for investors with money and foresight, it was the perfect time to invest in mining ventures. In the summer of 1925, a group of capitalists from Kansas City, Missouri -- later known as the Shenandoah-Dives Syndicate -- contracted mining engineer Charles A. Chase of Denver to travel to the San Juan mountains of southwestern Colorado to locate and purchase a gold mine for their investors.[1] Chase had first gone to the San Juan Mountains in 1899 at the invitation of Arthur Winslow, owner and manager of the Liberty Bell Mine in Telluride, Colorado. Working himself up from company surveyor and assayer, Chase eventually became general manager of the Liberty Bell Gold Mine. Chase spent 25 years with the Liberty Bell Mine until it was mined out in the 1920s.[2]

In August 1925, Chase traveled to Silverton to initiate preliminary exploration and development work in the area, beginning with an assessment of ore samples for gold and silver from the Old Hundred Gold Mine, which was located in a well-defined vein that ran through King Solomon Mountain. While collecting samples, Chase also assessed the rich veins of the Shenandoah-Dives and North Star mines. Chase's subsequent proposal to the Kansas City investors was to purchase and consolidate the Shenandoah-Dives, North Star, Terrible, and Mayflower mines into one large mine twice the size of the old Liberty Bell.[3]

Upon Chase's recommendation, the Kansas City group raised capital to purchase 31 patented claims and 12 unpatented claims covering 316 acres. In June 1929, under the laws of the State of Colorado, the syndicate incorporated as the Shenandoah-Dives Mining Company; James W. Oldham was the company president. The company's general offices were located in Kansas City, with mine operations near Silverton. The company was authorized to issue 3,500,000 shares of common stock at $1 par value.[4] In 1925-1927, the company began developing their Colorado holdings. Chase served as general manager of the Shenandoah-Dives mining operation. Prior to incorporation and in the development of the mine holdings, 27 men were employed as machine runners, trammers, cagers, and topmen; an engineer oversaw their work. They were assisted in non-mining enterprises by blacksmiths, carpenters, cooks, firemen and related helpers. Early production garnered $400,000. By July 1928, the nascent company also profited from the use of the Iowa Tiger Mill, a rented mill located in Arrastra Gulch.[5]

In 1928, Chase presented the syndicate with a prospectus outlining proposals for a tunnel, tram, and mill. The prospectus was accepted and funds transferred to Shenandoah-Dives Company and Charles Chase for construction to begin in June 1929. Arthur J. Weinig was hired as the designer of the metallurgical portion of the mining process. The Denver-based company of Stearns-Roger Engineering designed the structure that was to become the Shenandoah-Dives (Mayflower) Mill. Chase was in charge of the site and general layout.[6]

In addition to the mill complex, the Shenandoah-Dives Company developed the mine site, which was located on King Solomon Mountain. Here, the company built a boarding house for single miners, as well as those workers who did not live in Silverton.

(Company policy required single men to live in the boardinghouse.) The four-story boarding house contained storage rooms, heating plant, dining room, first-aid room, bakery and kitchen, offices, recreation room, commissary, staff quarters, and sleeping quarters. In addition to the boarding house, the above-ground buildings at the Mayflower Mine included a tramway terminal building, and a secondary tramway terminal to the boardinghouse. All the other facilities were built underground, due to the substantial snowfall and avalanches in the San Juan Mountains. These underground facilities included a crushing plant, compressed air plant, blacksmith shop, and foremen's office. The mine and the boarding house were connected via a short underground passage that allowed the men to work despite weather conditions.[7]

Transportation between the mine and the mill was by aerial tramway. Since 1870, mining operations in the Silverton area had relied on mules and aerial tramways for transportation, due to the high-altitude location of the mines and the treacherous mountain trails. The Shenandoah-Dives Company hired Fred C. Carstarphen, a well-known engineer in the region, to design the aerial tramway between the company's mine and the mill. Recognizing that snowfalls and avalanches often wiped out wooden tramways in the area, Carstarphen designed the

Shenandoah-Dives tramway with riveted steel towers. In addition, the Shenandoah-Dives' tramway's towers were larger than average. The Shenandoah-Dives aerial tramway was 9,526 feet long between the loading terminal and the discharge terminal, with a drop in elevation of approximately 1,400 feet. In 1930, this was believed to be the longest continuous aerial tramway in operation.[8] The tramway towers were built by the Pittsburgh Engineering Company, under the direction of Algot F. Andrean. Following construction, Andrean continued to work for the company as superintendent of the 12-man crew that maintained and operated the tramway. (In 1957, the Shenandoah-Dives discharge terminal tram building and aerial tramway were used as a movie set for Universal Studio's motion picture, Night Passage, with Jimmy Stewart. This is the only color-and-sound film of the tramway operating.)[9]

Due to the unique design of the system's three double-cable anchorages, the aerial tramway could withstand most avalanches. However, in 1938, an avalanche destroyed Towers 1 through 5. In an effort to prevent further avalanche damage, the company built an avalanche deflection structure above Towers 1 and 2 in the summer of 1938. The deflector was designed by Charles Chase, and built by Italian immigrant Carlo Poloine. Poloine

had come to Silverton from the Chicago area, where he had worked as a master stone mason on sewer and subway projects. Poloine was working as a timberman in the mine when Chase learned of his stone masonry skills and hired him to build the deflector.[10]

The avalanche deflector was built between May and September 1938. Algot F. Andrean, tramway supervisor, employed Tony Bazz, Eddie Valentine, Sam Mannick, and Joe Todeschi to assist Poloine in the construction of the avalanche deflector.[11] Poloine was paid $6.00 a day for an eight-hour day; while the others were paid $4.00 a day. Each morning, the men rode the tram up to the mine area where they met Poloine, who lived at the boarding house. Each man carried a bag of sand or cement up the hill about 150-200 feet to the site of the breaker. The men fashioned a wooden trail to the site to make hauling easier on the steep terrain. Water was carried up to the site to mix mortar, and for the men to drink. A mortar box was built at the site, and the deflector was constructed of rocks gathered from the mountainside.[12] When the deflector was completed, Chase declared it "a work of art." Afterwards, Poloine returned to being a timberman in the mine.[13] The deflector worked well to protect the upper terminal towers but did not save three lower towers from a devastating snowslide in 1963. Despite the loss of 1,500 feet of cable and towers, the double-cable anchorage system saved the rest of the aerial tramway from destruction.[14]

During the depression of the 1930s, many small, base-metals mining companies of the Rocky Mountains were not able to keep their operations afloat. However, the foresight of Charles A. Chase -- and the cooperative efforts of the town of Silverton and the Shenandoah-Dives Company -- enabled the mining and milling operations to survive the depression years.[15] During the 1930s, the Shenandoah-Dives Company's base metals (copper, lead, zinc) helped meet the needs of U.S. manufacturing companies, during a period when other regions suffered serious decline due to mine and mill foreclosures. While precious metals had been the reason for opening the Shenandoah-Dives mine and mill, base metals soon became the economic base of the company. By 1930, Shenandoah-Dives had invested $1,250,000 in the Silverton operations.[16] From 1930-32, the mill processed 461,826 tons of ore. At the time, Shenandoah-Dives was the largest, single, industrial payroll in the "Four Corners" area comprised of portions of Colorado, Utah, New Mexico, and Arizona.[17] By 1932 Chase also gained the company's workers cooperation in a

temporary pay reduction, in an effort to keep the mine open. Chase also was able to encourage businesses in the town to lower prices and accept credit from the miners to meet their basic needs; in turn, the miners' business kept the shops open. With the abandonment of the gold standard in 1933, the Silver Purchase Act of 1934, and the rise in gold prices from $20.67 to $35 per ounce, the mining regions of the West experienced a moderate boom period that ended the mining industry's depression. Miners' wages were returned to normal within two years and debts were settled.[18]

In 1938, 262 men worked for the Shenandoah-Dives Company in Silverton. By 1939, the wages of the employees of the Shenandoah-Dives operation were within the average of the industry. Miners, timbermen, trammers and loaders, trackmen, and motormen were paid $4.95 daily. The miners of Silverton formed the Silverton Miners Union, Local #26 of the Western Federation of Mine, Mill and Smelter Workers, in 1894.[19] The peak years of membership in Local #26 were from its inception in 1894 into the late 1920s, but unionization remained strong in the 1930s and 1940s in Silverton.[20] Unionization also had improved the miners' working conditions, including hours and wages. Congressional passage of the Hours-Wages Law in 1938 added strength to the union's battle for better working conditions. However, in August 1939, members of the Silverton Miner's Union, Local #26 became outraged when the Shenandoah-Dives Company lowered the wage base rate to counter the effects of the Hours-Wages Law enacted by Congress and required an eight-hour work day, overturning earlier agreements between union and company officials for a "portal to portal" workday of six hours.[21] The miners reacted by going on strike mid-July 1939.

After several months of negotiations between Chase, the Shenandoah-Dives Company, union negotiator A. S. Embree, and local union officers, union members decided to take matters into their own hands. The depression had closed many mining operations and the miners were anxious to resume operations before the Company closed Shenandoah-Dives. Members of the Local #26 disbanded their local to create a new local, representing their current concerns, under the San Juan Federation of Mines, Mills, & Smelter Workers. Assets of the disbanded Local #26 were turned over to the new union, and a new negotiations committee was formed to address workers' concerns with Charles Chase in order to resume work 7 September 1939.[22] As a result of the strike and following negotiations, Chase

implemented the contract mining system whereby workers who were more productive earned more. This, in turn, lowered company costs. Hoistmen and freighthandlers earned a daily wage of $4.40 and bucketmen earned $4.70, seven days a week; waiters received $80 per month, while cooks earned $165 per month.[23] Although Chase attempted to meet most of the needs of his employees, pressures from the Shenandoah-Dives Company's board of directors for a return on investment often forced him into an awkward and uncomfortable position of balancing the needs of the absent owners and the local workers.

Charles Chase's cost-efficient management also helped the Shenandoah-Dives operations remain viable. Chase gambled that the company's production of base metals, such as lead and zinc, would carry the cost of the operations, with additional profit from gold and silver. To insure the success of the Shenandoah-Dives operation, Chase built and ran the Silverton complex with the newest, most efficient mining and milling processes available. The Mayflower Mill was designed for the fullest recovery of metals, using alcohol reagents to separate three products from the ore. Upgrades in the mill in later years enabled the mill to separate five products.[24]

The Shenandoah-Dives Mill operation also reflected the increased environmental concerns of the mining and milling industry in the early 20th century. Since its territorial days, Colorado had laws outlawing stream pollution. Subsequent environmental laws were enacted, but were often ignored by mining companies. At the turn of the century, mining activities focused on recovering minerals from rock within which they were embedded in the quickest and most cost-efficient manner possible. Since it was generally accepted by society that mining contributed to the industrialization, modernization, and progress of the United States, the deleterious effects of mining on the surrounding environment or community often were not considered. However, this situation changed somewhat in 1884, when California's Yuba and Feather Valley farmers successfully sought retribution from hydraulic miners for environmental damage. This case set a precedent for further lawsuits by the agricultural community against the mining community. The California case also prompted the mining community to devise innovative mining processes conducive with the environment. While prior mining practices denuded timber growth, polluted water sources with tailings, and poured noxious gases into the air, Chase and a small number of other mine operators sought to install cleaner, more environmentally safe methods to help protect the locales in which they worked and lived.[25]

The Shenandoah-Dives operation was the first in the region to utilize tailings ponds. At the time, common practice was to slurry the tailings into available waterways, despite the dangers to the water supply and fish populations. Chase contacted J. T. Shimmin, who had devised a method of creating "ponds" of tailings at mills in Butte, Montana. After altering Shimmin's design to fit the triangular shape available at the Shenandoah-Dives Mill, and considering the soil content of the substructure, Chase had the mill's tailings deposited on the south side of the mill. As the pond was created, water was decanted off the surface, filtered and returned to the mill. A major part of the water evaporated or percolated through the tailings and into the substructure. The location of the ponds did not allow run off or percolation into the Animas River, which was east of the mill. After a trial period, with a few mishaps due to freeze and thaw, the tailing ponds of the Shenandoah-Dives Mill took shape in 1935. As one pond was filled to prescribed capacity, another pond was begun, ultimately resulting in four tailings ponds at the site.[26]

World War II ushered in a period of artificially inflated prices for essential resources, reviving the mining industry and the Shenandoah-Dives operations. As economic activity focused upon military preparedness, mining and milling operations were classified as either essential or non-essential. In 1942, the federal government suspended all non-essential mining activities for mines that counted more than 30% of their dollar value in gold and silver. The Shenandoah-Dives Mine and Mill's gold and silver production was a by-product from their base-metal ore production and was approximately the 30% legislative cap.[27] After a letter-writing campaign by Chase, the national publication, Mining World, and Colorado State legislators, the War Department granted the Shenandoah-Dives Mill permission to continue operations because its base metal (copper, lead, and zinc) production was deemed essential to national security. The Shenandoah-Dives 700-ton selective-flotation plant was one of the first plants to resume activity in the mining and milling community within the United States.[28]

By the early 1950s, however, the United States' metals prices were declining, due to the availability of cheaper foreign metals. During President Harry S Truman's administration, the Shenandoah-Dives Company received a government grant for the exploration of new veins, but lower grade ores and rising labor costs made any venture unprofitable for the company. Foreign metals flooded the United States market when the Paley Commission, appointed by President Dwight D. Eisenhower, encouraged the United States to purchase foreign metals to abate communist activity in smaller countries. As a result, metal prices collapsed, as well as the future of the Shenandoah-Dives operations in Silverton. As company officials debated the closure of the Shenandoah-Dives Mill, Charles A. Chase continued to fight for its survival. Chase organized a letter campaign to the Shenandoah-Dives Company's board of directors, Colorado congressmen, and banks. In letters touting the productivity of the plant and relating the artificial market of late, Chase sought money for exploration to keep the operation alive. But new investors and stockholders did not have the same emotional ties to the operation as Chase, and wanted quick profits.

In 1953, after 25 years of mining and milling in Silverton, the Shenandoah-Dives Company shut down its operations. During the previous 24 years, the mill had processed four million tons of Shenandoah-Dives ore, and 186,000 tons of custom ore from surrounding smaller enterprises, shipping the milled products to various smelters. In total, the

Shenandoah-Dives Mill had processed 11% of all the gold, silver, copper, lead, and zinc in Colorado. At the time, the assayed value of the Shenandoah-Dives Mill production was $32 million.[29] With the company's demise, hundreds of smaller enterprises were forced out of business.[30] A caretaker was retained for the mill property. Chase was fired as general manager and moved to Denver, Colorado, where he died on August 31, 1955.

Between 1953-1957, the mill operated only intermittently, as the Shenandoah-Dives Company underwent a series of ownership changes. Although the mill closed in 1953, the Shenandoah-Dives Company continued limited ore exploration in Silver Lake lease under the Defense Minerals Exploration Act (DMEA) grant. When no new ore was found under the DMEA grant, the Shenandoah-Dives Company sought to sell the mine and mill. In 1957, the Shenandoah-Dives Company merged with Marcy Exportation Corporation, a uranium mining company in Durango, Colorado and the mill reopened. The merger created the Marcy-Shenandoah Corporation.[31] Two years later, Marcy-Shenandoah Corporation sold its interests to Standard Uranium Corporation of Moab, Utah. In 1960, Standard Uranium Corporation changed its name to Standard Metals Corporation. The same year, Shenandoah-Dives Mine closed. The mill continued and processed ore from area mines. At the mill's peak, it employed ten men. In 1985, Standard Metals sold its Shenandoah-Dives holdings to the Sunnyside Gold Corporation, a subsidiary of the Echo Bay Mining Company of Edmonton, Canada. Sunnyside and associated companies participated in two joint ventures utilizing the mine and mill. By 1990, however, Sunnyside Gold was the sole owner again. In 1992, Sunnyside Gold announced permanent closure of the mine and mill, due to declining zinc prices and lack of gold reserves. By 1996, surface reclamation of the tailing ponds and mine site was substantially completed.[32] Following final closure, the population of Silverton declined by nearly one half of the remaining residents, as miners and their families migrated to other mining regions. As a result, San Juan County's tax base was reduced by one-third. Today, similar to other former mining towns in the Rocky Mountain West, the economy of Silverton has switched almost entirely to one based on government activities and tourism.[33]

Endnotes

1. William R. Jones, "Charles A. Chase & The Shenandoah-Dives Mine," Unpublished essay, 1994; Jones, correspondence to Dawn Bunyak, 1 April 1996; Margaret Barge, "History of the Shenandoah-Dives Mining Company," (TMs, 1959(?), San Juan County Historical Society, Silverton, Colorado, 3-6. Charles A. Chase, "Shenandoah-Dives Mine: Past, Present, and Future," San Juan County Historical Society Silverton, CO, 8 December 1954.

2. Ibid.

3. Ibid.

4. Jones, interview; and Barge, "History of Shenandoah," 3-6.

5. Barge, "History of Shenandoah," 8; Chase, "Past, Present;" Jones, interview; Report to the Bureau of Mines, State of Colorado for the year 1927: Metals, Mines, Mills, submitted by Shenandoah-Dives Syndicate for the Mayflower Mine, (County: San Juan; Dist.: Animas), 8 February 1928.

6. Jones, interview; Chase, "Past, Present;" and Report to the Bureau of Mines, State of Colorado for the years 1927 and 1929: Metals, Mines, and Mills, submitted by Shenandoah-Dives Syndicate for the Mayflower Mine, (County: San Juan; Dist.: Animas), 8 February 1928 and 28 May 1929.

7. Barge, "History of Shenandoah," 8 and Jones, correspondence to Dawn Bunyak 1 and 21 April 1996. The company paid $375,000 for the mill's construction and $85,000 for the aerial tramline.

8. Fred C. Carstarphen, "The Mayflower Aerial Tramway of the Shenandoah-Dives Mining Company," The Colorado School of Mines Magazine, XX, no. 9 (September 1930): 26 and "Shenandoah-Dives Part 2," Mining World, 5 (June 1942): 11 and 13.

9. William Jones, correspondence to Dawn Bunyak, 3 August 1995.

10. In a phone interview on December 16, 1996 between Joe Todeschi and Dawn Bunyak, Joe said Carlo Poloine was a middle-aged man, approximately 45 to 47 years old when they worked together on the building of the "snow breaker" above the number 1 and 2 towers of the Shenandoah-Dives Aerial Tramway.

11. During the 1996 interview with Dawn Bunyak, Joe Todeschi said that Tony Bazz was his step-father, who worked at the Shenandoah-Dives mine. Joe was never sure if he got the job because of his step-father or his persistence when approaching Al Andrean for a job after he had been laid-off. "One day I collared Al on the street and asked him for a job, as I had gotten laid off at Sunnyside where I had been working on the installation of their tram," Joe related. "Mr. Andrean, I need me a job real bad," I said. "I had to eat you know." Continuing his conversation with Mr. Andrean, Joe said, "Do you have anything for me?" Mr. Andrean was obliged to tell Joe that he had no work then but would keep the boy in mind. "The next day, Mr. Andrean came to our door and told me he had some work for me," Joe continued.

12. Todeschi, interview.

13. Ibid.

14. "Shenandoah-Dives," Mining World, 7; Jones, interview; and Jones, correspondence, 3 August and 8 August 1995.

15. "Report to the Bureau of Mines, State of Colorado for the years 1929-1938: Metals, Mines, and Mills," submitted by Shenandoah-Dives Syndicate for the Mayflower Mine, (County: San Juan; Dist.: Animas), 28 May 1929, 6 December 1929, 15 August 1930, and 5 January 1938.

16. Barge, "History of Shenandoah-Dives," 4.

17. William Jones, interview by Dawn Bunyak, Silverton, CO, 15 June 1995.

18. Beverly Rich, Chair of San Juan County Historical Society, correspondence to Dawn Bunyak on 6 November 1995. Ms. Rich's research on the Silverton Miners Union, Local #26, found that the union representatives and company officials had agreed on a six hour workday for miners citing hazardous and hard working conditions. This "portal to portal" exception came after many years of negotiations.

19. Beverly Rich, Chair of San Juan County Historical Society, correspondence to Dawn Bunyak on 6 November 1995.

20. Ibid.

21. "San Juan Federation of Mine, Mill and Smelter Workers Formed," The Silverton Standard, 1 Sept. 1939.

22. Ibid.

23. Barge, "Shenandoah;" Jones, "Chase;" and "Report to the Bureau of Mines, State of Colorado for the years 1929-1938: Metals, Mines, and Mills," submitted by Shenandoah-Dives Syndicate for the Mayflower Mine, (County: San Juan; Dist.: Animas), 28 May 1929, 6 December 1929, 15 August 1930, and 5 January 1938.

24. Chase and Kentro, "Tailings Disposal," 1937; and Chase and Kentro, "Tailings Disposal," 1938.

25. Carl Ubbelohde, Maxine Benson, Duane A. Smith, A Colorado History, 6 ed., (Boulder, CO: Pruett Publishing Co., 1988), 333-36, 346.

26. Chase and Kentro, "Tailings Disposal, 1937," 20-22; and Chase and Kentro, "Tailings Disposal, 1938," 31-32, 74.

27. "Shenandoah-Dives Part 2," Mining World, 10.

28. Carl Ubbelohde, Maxine Benson, Duane A. Smith, A Colorado History, 6 ed., (Boulder, CO: Pruett Publishing Co., 1988), 333-36, 346; and "Shenandoah-Dives, Part 2," Mining World, 10.

29. "Mine, Mill Shutdown Held a 'Catastrophe'," Denver Post, 19 February 1953, p. 60; Barge, "History of Shenandoah," 11; and Jones, "Chase."

30. Barge, "History of Shenandoah," 11; Jones, "Chase;" "Mine, Mill Shutdown Held a 'Catastrophe'," Denver Post, 19 February 1953, 60; "End of a Chapter," editorial, Denver Post, 21 February 1953; and Obituary of Charles A. Chase, Rocky Mountain News, 1 Sept. 1955, 14.

31. William Jones, correspondence to Dawn Bunyak, 1 April 1996.

32. William Jones, correspondence to Dawn Bunyak, 1 April 1996 and Darnall Zanoni, correspondence to Dawn Bunyak, 1 April 1996.

33. Beverly Rich, San Juan County Treasurer and Chairperson of the San Juan County Historical Society, correspondence to Dawn Bunyak on 6 November 1995. Ms. Rich's interest in the historical preservation of her community has motivated her to research numerous topics in local records concerning the Miner's Union #26, San Juan Federation of Mine, Mill, & Smelter, Miner's Union Hospital, and mine labor issues. She has received grant funding from the Colorado Historical Society for her research in labor issues.

A Sampling of Flotation Mills in the United States

As I began to search for flotation mills in 1997, I accumulated a list of mills that were either extant, deteriorating, ruins, or removed. Most of the flotation mills that I heard or read about during the research stage are listed below. As the list began to grow and time constraints were applied, the objective of the report was tightened to include only the location of extant 20th-century flotation mills. Therefore, this list must be looked upon as only a partial listing of U.S. flotation mills. Many mining districts have ruins of mills, but it is not always possible to determine if they were flotation mills, rather than stamp, cyanide, or combination mills. To the best of my knowledge, the following sites contained flotation mills. In the event of mistakes, I bear responsibility.

Independence Mill	near Anchorage, AK
Juneau District	near Juneau, AK
Cliff Mine and Mill	Prince William Sound, AK
Port Wells	Prince William Sound, AK
Ellamar	Prince William Sound, AK
New Cornelia Plant	Ajo, AZ
Shattuck-Denn Mill	Bisbee District, AZ
Hillside Mine & Mill	Eureka District, Yavapai County, AZ
Bagdad Mine & Mill	Eureka District, Yavapai County, AZ
Globe Miami Mill	Gila County, AZ
Clifton-Morenci Plant	Greenlee County, AZ
Miami Concentrator	Globe, AZ
Old Dominion	Globe, AZ
United Verde Mill	Jerome District, Yavapai County, AZ
Detroit Copper	Morenci, AZ
Phelps Dodge Corp, Mill	Morenci, AZ
Kennecott Corp. Plant	Pinal County, AZ
Crown King Mine & Mill	Pine Grove-Tiger District, AZ
Flotation Mill	Poland, AZ
Magma Mill	Superior, AZ
Bodie Mill	Bodie State Park, CA
Shattuck Denn Mining	Bisbee, CA
Globe-Miami Mill	Gila County, CA
Clifton-Morenci Mill	Greenlee, CA
Engels Copper	Taylorsville, CA
Climax Mill	Climax, CO
Solomon Mill	Creede, CO
Creede Mill	Creede, CO
Economic Extraction Mill	Cripple Creek, CO
Carlton Mill	Cripple Creek, CO

Sunnyside Mill	Eureka, CO
Argo Mill	Idaho Springs, CO
Mary Murphy Mine & Mill	Iron City, CO
Montezuma Mill	Montezuma, CO
Idarado Mill	Telluride, CO
Independence Mill	Victor, CO
Golden Cycle Mill	Victor, CO
Portland Mill	Victor, CO
Eureka Mill	CO
Graham-Snyder Mining & Mill	Galena, IL
North Star Mill	Hailey, ID
Bunker Hill and Sullivan Concentrator	Kellog, ID
Sunshine Mill	Kellog, ID
Morning Mill	Mullen, ID
Calumet & Hecla Consolidated Mill	Houghton, MI
St. Joseph Lead Mill	Bonne Terre, MO
Selective Flotation Mill at Mill Pond	Grandin, MO
Herculaneum Mill	Herculaneum, MO
Anaconda Concentrator	Anaconda, MT
Anaconda Concentrator	Anaconda, MT
Bannack Mill	Bannack, MT
Comet Town Site	Basin, MT
Kennecott Mill	Butte, MT
Timber Butte Mill	Butte, MT
Helena Mill	Helena, MT
Royal Basin Mine & Mill	Maxville, MT
Slowey Mill	Mineral County, MT
Nancy Lee Mill	Superior, MT
General Custer Mill	MT
Reed Gold Mill	NC
Socorro Mill	Mogollon, NM
Santa Rita Copper Mill	NM
Venadium Mill	NM
Combined Metals Mill	Castleton, NV
Florence Mine & Mill	Goldfield, NV
Consolidated Nevada & Utah Corp Mill	Pioche, NV
Comstock Mill	near Reno, NV
Belmont Mine & Mill	Tonopah, NV
Carlontrend/ Carlon Trend Mill	NV
Homestake Mill	SD
Burra Burra Mine & Mill	Ducktown District, TN
Utah-Apex Mining	Bingham, UT
Magna & Arthur Plants	Garfield, UT
Midvale Plant	Midvale, UT

Utah Leasing Plant	Newhouse, UT
Silver King Coalition Mine & Mill	Park City, UT
Utah Consolidated Mill	Park City, UT
Chief Consolidated Mill	Tintic District, UT
Tintic Standard Reduction Mill	Tintic District, UT
Tooele Mill	Tooele, UT
Knob Hill Mill	Knob Hill, WA
Mill	Metalline, WA
National Roaster Plant	Cuba City, WI
Rule Mining Co. Mill	Lafayette County, WI
New Jersey Zinc Plant	Platteville, WI
Vinegar Hill Plant	Shullsburg, WI
Eagle Picher Plant	Shullsburg, WI

Works Cited

Primary Sources

Manuscript Collections

Barge, Margaret. "History of the Shenandoah-Dives Mining Company." Manuscript, 1959(?). San Juan County Historical Society, Silverton, Colorado.

Chase, Charles A.. Collection [photostat] in private collection of William Jones, Montrose, Colorado.

Chase, Charles A.. "Shenandoah-Dives Mine: Past, Present, and Future." Unpublished essay. 8 December 1954. Included in Ms. by Margaret Barge. San Juan County Historical Society Archives, Silverton, Colorado.

Jones, William, Vice-President of Shenandoah Mining Company, Inc. Montrose, Colorado. "Charles A. Chase & The Shenandoah-Dives Mine." Unpublished essay, 1994. Photostat of original, National Park Service-Intermountain Support Office, Lakewood, Colorado.

Unknown. "Western Colorado Power Company History." Unpublished manuscript, 1979(?). San Juan County Historical Society Archives, Silverton, Colorado. Chapters 4, 8, and 13.

Government Documents

Colorado Historical Society. Historic Building Inventory. Photostat. Mayflower Mill. 1995.

"Report to the Bureau of Mines, State of Colorado for the years of 1927, 1929, 1930, 1937, 1949, 1955, 1958, 1962, 1964, 1973, and 1975: Metals, Mines, and Mills." Submitted by Shenandoah-Dives Syndicate for the Mayflower Mine, (County: San Juan, Dist.: Animas), 1928, 28 May 1929, 6 December 1929, 15 August 1930, 5 January 1938, 8 February 1950, 4 November 1955, 2 October 1958, 21 January 1963, 9 November 1964, 25 October 1973, and 15 December 1975. (Photocopy in the private collection of William Jones, Montrose, Colorado.)

"Report to the Bureau of Mines, State of Colorado for the year 1960: Operators Annual Report." Submitted by Shenandoah-Dives Syndicate for the Mayflower Mine, (County: San Juan, Dist.: Animas), 8 February 1961. (Photocopy in the private collection of William Jones, Montrose, Colorado.)

Thrush, Paul W. A Dictionary of Mining, Mineral, and Related Terms. U.S. Department of the Interior. Geological Survey. Washington, 1968.

U.S. Department of the Interior. Bureau of Land Management. Mineral Survey 7608. Field notes and Plat of the Polar Star Mill Site. San Juan County, Colorado. 1 August 1892. Lakewood, Colorado.

--------. Geological Survey. Principal Gold-Producing Districts of the United States, by A.H. Koschmann and M.H. Bergendahl. Washington, D.C.: U.S. Government Printing Office, 1968.

--------. Division of National Register Programs. Death Valley to Deadwood Kennecott to Cripple Creek: Proceedings of the Historic Mining Conference January 23-27, 1989 in Death Valley Monument, ed. Leo R. Barker and Ann E. Huston. San Francisco: Government Printing Office, 1990.

--------. National Park Service. Division of National Register Programs. Guidelines for Identifying, Evaluating, and Registering Historic Mining Properties, by Bruce J. Noble and Robert Spude. National Register Bulletin 42. Washington, D.C.: Government Printing Office, 1992.

--------. National Park Service. Division of National Register Programs. How to Apply the National Register Criteria for Evaluation. National Register Bulletin 15. Washington, D.C.: Government Printing Office, 1992.

--------. National Park Service. Division of National Register Programs. How to Complete the National Register Registration Form. National Register Bulletin 16A. Washington, D.C.: Government Printing Office, 1992.

--------. National Park Service. Historic American Buildings Survey and the Historic American Engineering Record. Washington, D.C.

--------. National Park Service. The National Survey of Historic Sites and Buildings: The Mining Frontier by Dr. Benjamin F. Gilbert, ed. by William C. Everhart. Washington, D. C.: Government Printing Office, 1959.

Whitacre, Christine, Front Range Research, Inc. National Park Service, Denver, CO. "Telluride National Historic Landmark District." National Register of Historic Places Registration Form, 30 September 1988.

Miscellaneous Primary Sources

Carstarphen, Fred C. "The Mayflower Aerial Tramway of the Shenandoah-Dives Mining Company." The Colorado School of Mines Magazine XX, no. 9 (September 1930): 24-26.

--------. "Flotation Widely Employed in San Juan Milling." Engineering and Mining Journal 136, no. 8 (August 1935): 395-397.

--------. "Widely Known Colorado Mining Engineer Dies." The Rocky Mountain News (Denver), 1 September 1955, 14.

Chase, Charles A. and Dan Kentro. "Tailings Disposal Practice of Shenandoah-Dives Mining Company." In Mining Year Book of Intermountain Region, Colorado Chapter of the American Mining Congress, 31-32, 74-75. Denver, CO: Colorado Mining Association, 1937.

--------. "Tailings Disposal Practice of Shenandoah-Dives Mining Company." Mining Congress Journal 24 (March 1938): 19-22.

Durell, C. Terry. "Universal Flotation Theory." Colorado School of Mines Magazine 6 (February 1916): 27-34.

"End of a Chapter." Editorial. Denver Post, 21 February 1953.

(Helena) Montana Record-Herald, 12 July 1939, sec. 5, 3.

"The History of the Flotation Process." The Mining Magazine vol. 1 (September 1909): 61-64.

Hunt, T. R., and Leon M. Banks. "Operations of the Shenandoah-Dives Mining Company." The Mines Magazine (February 1934): 46-52.

"Mine, Mill Shutdown Held a 'Catastrophe'." Denver Post, 19 February 1953, 60.

Mining Directory of San Miguel, Ouray, San Juan and La Plata Counties. Denver: County Directory Co., Jan. 1899.

"Polar Star Mill." Denver Republican, 12 January 1882, 6.

San Juan County, Commemorative Issue. Silverton Standard, Colorado. 1 December 1899.

"San Juan Federation of Mine, Mill and Smelter Workers Formed." Silverton Colorado Standard. 1 September 1939.

"Shenandoah-Dives." Mining World 4 (April 1942): 3-7.

"Shenandoah-Dives. Part Two." Mining World 5 (June 1942): 10-13.

The Silverton (Colorado) Miner. Golden San Juan Ed. 1907: 9-ll. Collection of William Jones of Montrose, Colorado.

"The Trend of Flotation." The Colorado School of Mines Magazine 25 (January 1930): 20.

Interviews

Affleck, Carol. Phone interview by Dawn Bunyak. 23 & 24 January 1997.

Athearn, Frederic J., Historian. Bureau of Land Management, Lakewood, Colorado. Phone interview by Dawn Bunyak. 6 September 1995.

Beuchler, Jeffrey, Consultant. Dakota Research Service. Rapid City, SD. Phone interview by Dawn Bunyak. 11 March 1997.

Bedeau, Burt, Archivist and Historian. National Register Office, Idaho State Historical Society, Boise, ID. Phone interview by Dawn Bunyak. 24 February 1997.

Cuneo, Kurt, Forest Ranger. Helena Ranger District, Helena, MT. Phone interview by Dawn Bunyak. 5 March 1997.

Dean, Glenna, Archeologist. Mining and Minerals Division, Santa Fe, NM. Phone interview by Dawn Bunyak. 13 March 1997.

DeLony, Eric, Chief of the Historic American Engineering Record. Washington, D.C.. Phone interview by Dawn Bunyak. 4 March 1997.

Fuller, Willard P., Retired Mining Engineer and Geologist. San Andreas, CA. Phone interview by Dawn Bunyak. 4 March 1997.

Graeme, Richard, Mining Engineer, Golden Queen Mining. Mojave, CA. Phone interview by Dawn Bunyak. 28 March, 16 April, and 10 July 1997.

Horstman, Mary, Historian for Region One, U. S. Forest Service--Bittersweet District. Hamilton, MT. Phone interview with Dawn Bunyak. 25 February 1997.

Jensen, Bruce, State Historian. National Register Program, Texas Historical Commission, Austin, TX. Phone interview by Dawn Bunyak. 5 March 1997.

Jones, William, Vice-President of Shenandoah Mining Company, Inc. Montrose, Colorado. Interview by Dawn Bunyak. 15 June 1995.

--------. Correspondence to Dawn Bunyak. 3 August 1995.

--------. Correspondence to Dawn Bunyak. 8 August 1995.

--------. Phone interview with Dawn Bunyak. 7 January 1997.

Kaldenberg, Russ, Archeologist. California Regional Office, Bureau of Land Management, Sacremento, CA. Phone interview by Dawn Bunyak. 4 March 1997.

Knapp, Reed, Historian. Raleigh, NC. Phone interview by Dawn Bunyak. 6 March 1997.

Koreth, John, Asst. Bureau Chief. Dept. of Environmental Quality, Helena, MT. Phone interview by Dawn Bunyak. 25 February 1997.

Krabaucher, Paul, Project Manager of the Western Slope. Colorado State Office of Mining, Denver, Colorado. Phone interview by Dawn Bunyak. 11 September 1996.

Langenfeld, Mark, Mining historian. Madison, WI. Email correspondence to Dawn Bunyak. 6 February 1997.

McCulloch, Robin, Mining Engineer, Montana Bureau of Mines and Geology. Butte, MT. Phone interview by Dawn Bunyak. 26 March 1997.

Nossaman, Allen, San Juan County Judge. Interview by Dawn Bunyak. 13-15 June 1995.

Peterson, Freda, Cemetery Historian. Interview by Dawn Bunyak. 14 June 1995.

Philips, Kenneth, Chief Engineer. Department of Mines and Mineral Resources, Phoenix, AZ. Phone interview by Dawn Bunyak. 5 March 1997.

Portfleet, Matthew G. Michigan Technological University, Houghton, MI. Email correspondence to Dawn Bunyak. 6 February 1997.

Rich, Beverly, Chair of San Juan County Historical Society, Silverton, CO. Phone interview by Dawn Bunyak. 20 August 1995.

--------. Correspondence to Dawn Bunyak. 6 November 1995.

Roth, Susan, State Historian. National Register Program, Minnesota Historical Society, St. Paul, MN. Phone interview by Dawn Bunyak. 11 March 1997.

Smith, Duane, Historian and Professor. Fort Lewis College, Durango, CO. Phone interview by Dawn Bunyak. 7 January 1997.

Spude, Robert, Program Leader, National Register Programs. Southwest Support Office, National Park Service, Santa Fe, NM. Interview by Dawn Bunyak. 6 July 1995.

--------. Email correspondence to Dawn Bunyak. 24 February 1997.

--------. Phone interview with Dawn Bunyak. 6 January 1997.

Toms, Donald, Curator. Black Hills Mining Museum, Lead, SD. Phone interview by Dawn Bunyak. 11 March 1997.

Wegman-French, Lysa, Historian. Rocky Mountain Support Office, National Park Service, Denver, CO. Email correspondence to Dawn Bunyak. 13 February 1997.

Todeschi, Joseph, Sr., Retired Miner. Phone interview with Dawn Bunyak. 16 December 1996.

Zanoni, Darnall, Retired Miner. Interview by Dawn Bunyak. 13-15 June 1995.

Secondary Sources

Arbiter, Nathaniel, ed. Milling Methods in the Americas. New York: Gordon and Breach Science Publishers, 1964.

Bailey, Ronald H.. The Homefront U. S. A. New Jersey: Time-Life Books, Inc., 1978.

Burt, Richard O. and Chris Mills. Gravity Concentration Techniques. Amsterdam: Elsevier, 1984.

Cobb, Harrison. Prospecting Our Past: Gold, Silver, and Tungsten Mills of Boulder County. Boulder, CO: The Book Lode, 1988.

Crabtree, E. H. and J. D. Vincent. "Historical Outline of Major Flotation Developments." In Froth Flotation 50th Anniversary ed. D. W. Fuerstenau. New York: American Institute of Mining, Metallurgical and Petroleum Engineering Inc., 1962.

Dorenfeld, Adrian C. "Flotation Circuit Design." In Froth Flotation 50th Anniversary. ed. D. W. Fuerstenau. New York: American Institute of Mining, Metallurgical and Petroleum Engineering Inc., 1962.

Dunbar, A. R. Western Mining Directory 1901-1902. Denver & San Francisco: Western Mining Directory Co., 1901.

Flotation Fundamentals and Mining Chemicals. Midland, MI: Dow Chemical Co., 1968.

Francaviglia, Richard V. Hard Places: Reading the Landscape of America's Historic Mining Districts. Iowa City, IA: University of Iowa Press, 1991.

Fuerstenau, D. W., ed. Froth Flotation 50th Anniversary. New York: American Institute of Mining, Metallurgical and Petroleum Engineering Inc., 1962.

Gaudin, A. M. Flotation. New York: McGraw-Hill Book Co., Inc., 1932.

--------. Principles of Mineral Dressing. New York and London: McGraw-Hill Book Co., 1967.

Greenough, W. Earl. First Hundred Years of the Coeur D'Alene Mining Region. Mullan, ID: American Institute of Mining and Metallurgical Engineering, 1947.

Gregory, Cedric E. A Concise History of Mining. New York: Pergamon Press, 1980.

Hardesty, Don L. The Archaeology of Mining and Miners: A View from the Silver State. Special Publication Series, no. 6. Ann Arbor, Michigan: The Society for Historical Archaeology, 1988.

Hines, Pierre R. "Before Flotation." In Froth Flotation 50th Anniversary Volume. ed. D. W. Fuersteanu. New York: American Institute of Mining, Metallurgical and Petroleum Engineering Inc., 1962.

Hoover, Theodore J. Concentrating Ores by Flotation. San Francisco: Mining and Scientific Press, 1914.

The New Encyclopaedia Britannica. Chicago: Encyclopaedia Britannica, Inc., 1993.

Mouat, Jeremy. "The Development of the Flotation Process: Technological Change and the Genesis of Modern Mining, 1898-1911." Australian Economic History Review 36, No. 1 (March 1996): 3-31.

Peele, Robert. Mining Engineers' Handbook. 2 vol. New York: John Wiley & Sons, Inc., 1918, 1927, 1941.

Perrett, Geoffrey. Days of Sadness, Years of Triumph: The American People 1939-1945. New York: Coward, McCann and Geoghegan, Inc., 1973.

Richards, Robert. Ore Dressing. New York: McGraw-Hill Book Co., 1909.

--------. Textbook of Ore Dressing. New York & London: McGraw-Hill Book Co., Inc., 1940.

Rickard, Thomas A. A History of American Mining. New York & London: McGraw-Hill Book Co., Inc., 1932.

--------. Concentration by Flotation. New York: John Wiley and Sons, Inc., 1921.

Rickard, Thomas A. and O. C. Ralston. Flotation. San Francisco: Mining and Scientific Press, 1917.

Schlesinger, Arthur M., editor. Almanac of American History. New York: Barnes and Noble Books, 1993.

Sloane, Howard N. & Lucille L. A Pictorial History of American Mining. New York: Crown Pub., Inc., 1970.

Works Cited

Smith, Duane A. Mining America: The Industry and the Environment, 1800-1980. Kansas City: University of Kansas Press, 1987.

Ubbelohde, Carl, Maxine Benson, and Duane A. Smith. A Colorado History 6th ed. Boulder, CO: Pruett Pub. Co., 1988.

Wegman-French, Lysa. "The History of the Holden-Marolt Site in Aspen, Colorado: The Holden Lixiviation Works, Farming and Ranching, and the Marolt Ranch 1879-1986" Aspen, CO: Aspen Historical Society, October 1990.

Wolle, Muriel S. Stampede to Timberline: The Ghost Towns & Mining Camps of Colorado. Boulder, CO: By the author, 1949.

*U.S. GOVERNMENT PRINTING OFFICE 1999-0-773-551